AF323008

# Approximation
## with
## Quasi-Splines

Also by IOP Publishing

Scientific Programmer's Toolkit

*M H Beilby et al*

A Simple Introduction to Numerical Analysis

*R D Harding and D A Quinney*

Microcomputer Algorithms

*J P Killingbeck*

Compact Numerical Methods for Computers

*J C Nash*

Walsh Series
An Introduction to the Dyadic Harmonic Analysis

*F Schipp et al*

# Approximation with Quasi-Splines

**G H KIROV**
Plovdiv Higher Institute of
Food and Flavour Industries, Bulgaria

Institute of Physics Publishing
Bristol, Philadelphia and New York

*British Library Cataloguing-in-Publication Data*

Kirov, G. H.
  Approximation with quasi-splines.
I. Title
515
ISBN 0-7503-0181-3

*Library of Congress Cataloging-in-Publication Data are available*


Published by IOP Publishing Ltd, a company wholly owned by the Institute of Physics, London

IOP Publishing Ltd
Techno House, Redcliffe Way, Bristol BS1 6NX, England
335 East 45th Street, New York, NY 10017-3483, USA
US Editorial Office: 1411 Walnut Street, Suite 200, Philadelphia, PA 19102


Typeset by P & R Typesetters Ltd, Salisbury, Wilts
Printed in Great Britain by Galliard (Printers) Ltd, Great Yarmouth

# Contents

# Preface

The present book is an attempt to summarize a series of investigations on the spline approximations initiated by Favard, Chakalov and Schoenberg, and continued by Korneichuk and Malozemov *et al*. In a sense, it reflects the personal preferences of the author, but nevertheless I believe that it may be useful for other people also.

As is well known, the solutions of many extremal problems of recovery of functions and functionals are given by spline-functions. So, it is interesting to look for other classes of functions with similar approximative properties to the spline functions. As a generalization of the spline functions of the first degree, the basic quasi-splines are introduced. In this connection, the relation of the quasi-splines with the atomar function is of special interest. But there exists a threshold in the approximation of functions by quasi-splines. This is a reason for introducing the notion of the fundamental rational quasi-spline (in particular, the rational spline). In this book its approximative properties are studied.

I have the pleasant duty of expressing my gratitude to Professor Ivan Dimovski for proposing that this book should appear in English, for continuous support, and for help in translating the Bulgarian text into English. I am aware that without the help of Dr M Petkov from the Institute of Mathematics of the Bulgarian Academy of Sciences and that of Mrs S Stefanova from Jus Autor, Sofia, this book might well have not appeared as an English edition. I express my deep gratitude to them. Last, but not least, I wish to thank the copy editor of the book Neil Scriven for his excellent, highly professional work on the manuscript.

**Georgy H Kirov**
18 April 1991
Plovdiv

# 1

# Introduction

## 1.1  Motivation

In this work the common notation $\mathbb{N}$, $\mathbb{Z}$, $\mathbb{Q}$ and $\mathbb{R}$ is used for the set of positive integers, integers, rational numbers and real numbers, respectively. The notation $\mathbb{N}_0 = \mathbb{N} \cup \{0\}$ and $\mathbb{R}_+$ is also used.

Any system $x_1$, $x_2$, ..., $x_{n-1}$, $x_n$ of $n (n \in \mathbb{N})$ different points from the segment $[0, 1]$, arranged in increasing order

$$0 \leqq x_1 < x_2 < x_3 < \ldots < x_n \leqq 1 \tag{1.1.1}$$

is said to be a mesh with nodes $x_1$, $x_2$, $x_3$, ..., $x_{n-1}$, $x_n$. We denote it by $X = (x_1, x_2, \ldots, x_n)$.

Any system

$$k_1(x), \qquad k_2(x), \ldots, k_n(x), \qquad x \in [0, 1] \tag{1.1.2}$$

of $n$ nonnegative functions which form a partition of the unit:

$$k_i(x) \geqq 0 \quad (i = 1, 2, \ldots, n), \qquad \sum_{i=1}^{n} k_i(x) = 1, \qquad x \in [0, 1] \tag{1.1.3}$$

is said said to be a system of basic (or fundamental) quasi-splines.

Let the function $f : [0, 1] \to \mathbb{R}$ have derivatives up to $r$th order ($r \in \mathbb{N}_0$, $f^{(0)}(x) = f(x)$) at each point $x \in [0, 1]$. The Taylor polynomial of order $r$ of $f(x)$ in a neighbourhood of a point $\alpha \in [0, 1]$ will be denoted by

$$S_r(f, \alpha; x) = \sum_{k=0}^{r} \frac{f^{(k)}(\alpha)}{k!} (x - \alpha)^k.$$

For each such function $f$ we will suppose that the values $f^{(k)}(x_i)$ of the function and its derivatives ($k = 0, 1, \ldots, r$; $i = 1, 2, \ldots, n$) are known (or, can be calculated) up to $r$th order at the nodes of the mesh $X = (x_1, x_2, \ldots, x_n)$, i.e. we suppose that we know the matrix

$$T_r(f, X) = \left\| \begin{matrix} f(x_1) & f(x_2) & \ldots & f(x_n) \\ f'(x_1) & f'(x_2) & \ldots & f'(x_n) \\ . & . & \ldots & . \\ f^{(r)}(x_1) & f^{(r)}(x_2) & \ldots & f^{(r)}(x_n) \end{matrix} \right\|. \tag{1.1.4}$$

It is said to be the information for the function $f$ on the mesh $X$.

The quasi-spline of order $r$ generated by $(T_r(f, X), k_i(x))$ is the function

$$\Phi(x) = \Phi_n^r(x) = \Phi(k, T_r(f, X); x) = \Phi^r(f, X; x)$$

$$= \sum_{i=1}^{n} k_i(x) S_r(f, x_i; x) = \sum_{i=1}^{n} \sum_{k=0}^{r} k_i(x) \frac{f^{(k)}(x_i)}{k!} (x - x_i)^k \tag{1.1.5}$$

of $x \in [0, 1]$.

We shall study the approximative properties of the quasi-splines of order $r$ in various classes of functions under different measures of proximity under appropriate choices of the fundamental quasi-splines.

The motivations of this problem are the following:
(a) the quasi-splines of order $r(r \in \mathbb{N}_0)$ are the solutions of extremal problems arising in practice;
(b) for given information $T_r(f, X)$ for a function $f$ on a mesh $X$ the values of $f$ in each point of its domain can be calculated to an arbitrary accuracy;
(c) the quasi-splines contain various spline-functions of interpolation type, and also the atomar functions. The properties of the atomar functions can be obtained as a special case of the properties of the quasi-splines.


## 1.2   Moduli of continuity

We recall the definitions and properties of some moduli of continuity (see, e.g. [1] pp 17–20, [2], pp 11–34).

Let $C[a, b](C[0, 1] = C)$ be the 'normed' linear space of all continuous functions $f : [a, b] \to \mathbb{R}$ with the norm

$$\| f \|_C = \| f \|_{C[a,b]} = \sup\{|f(x)| : x \in [a, b]\}$$

The modulus of continuity of a bounded function $f : [a, b] \to \mathbb{R}$ is said to be the following function of $\delta \in [0, b - a]$

$$\omega(f; \delta) = \sup\{|f(x) - f(x')| : |x - x'| \leq \delta; x, x' \in [a, b]\}. \tag{1.2.1}$$

The function $\omega(f;\delta)$ has the following properties

(i) $\omega(f;0)=0$;

(ii) $\omega(f;\delta)$ is increasing on the segment $[0,b-a]$;

(iii) $\omega(f;\delta)$ is a semi-additive function of $\delta\in[0,b-a]$,

    i.e. $\omega(f;\delta+\delta')\leqq\omega(f;\delta)+\omega(f;\delta')$,    $\delta,\delta',\delta+\delta'\in[0,b-a]$;

(iv) $\lim\limits_{\substack{\delta\to0\\(\delta>0)}}\omega(f;\delta)=\omega(f;0)=0\Leftrightarrow f\in C[a,b]$.

From (iii) and (iv) follow

(v) $\omega(f;\cdot)\in C[0,b-a]\Leftrightarrow f\in C[a,b]$.

These properties are characteristic of the modulus of continuity in the following sense: each function $\omega(\delta)$, $\delta\in[0,\tau]$ which is increasing, semi-additive and continuous on the segment $[0,\tau]$ and vanishing at $0\,(\omega(0)=0)$ coincides with its modulus of continuity: $\omega(\omega;\delta)=\omega(\delta)$, $\delta\in[0,\tau]$.

This fact allows us to propose a definition of the notion of modulus of continuity, independent of the specific function $f$.

---

**Definition 1.2.1**    A *modulus of continuity* is said to be any function

$$\omega(\delta),\,\delta\in[0,+\infty)\text{ or }\delta\in[0,\tau],\,\tau>0$$

which is increasing, semi-additive, continuous, and with $\omega(0)=0$.

---

The function

$$\omega(t)=t^{\alpha},\qquad0<\alpha\leqq1,\qquad t\in[0,+\infty]\qquad(1.2.2)$$

is an important example of a modulus of continuity.

A generalization of the moduli of continuity are the moduli of smoothness.

For an arbitrary bounded function $f:[a,b]\to\mathbb{R}$ and for each fixed $k\,(k\in\mathbb{N})$, the $k$th modulus of smoothness for the function $f$ is said to be the following function of $\delta\in[0,(b-a)/k]$:

$$\omega_k(f;\delta)=\sup\{|\Delta_h^k f(x)|:|h|\leqq\delta;\,x,x+hk\in[a,b]\}\qquad(1.2.3)$$

with

$$\Delta_h^k f(x)=\sum_{i=0}^{k}(-1)^{k+i}\binom{k}{i}f(x+ih),$$

$$\Delta_h f(x)=\Delta_h^1 f(x),\qquad\binom{k}{i}=\frac{k!}{i!(k-i)!}.$$

The second modulus of smoothness $\omega_2(f;\delta)$ is also called Zygmund's modulus for the function $f$.

Obviously, $\omega_1(f;\delta) = \omega(f;\delta)$.

The moduli of smoothness have the following basic properties ([2], pp. 12–13)

(i) $\omega_k(f;\delta)$ is increasing in $[0,(b-a)/k]$;

(ii) $\omega_k(f+g;\delta) \leqq \omega_k(f;\delta) + \omega_k(g;\delta)$—semi-additivity with respect to the function;

(iii) $\omega_{k+1}(f;\delta) \leq 2\omega_k(f;\delta)$—estimating the $(k+1)$th modulus by means of the $k$th modulus;

(iv) $\omega_{k+1}(f;\delta) \leq \delta\omega_k(f';\delta)$—estimating the $(k+1)$th modulus by means of the $k$th modulus of the derivative;

(v) $\omega_k(f;n\delta) \leq n^k\omega_k(f;\delta)$—homogeneity with respect to the positive integer multiplier $n\in\mathbb{N}$;

(vi) $\omega_k(f;\lambda\delta) \leq (\lambda+1)^k\omega_k(f;\delta)$—homogeneity with respect to a positive multiplier;

(vii) If the function $f$ has a bounded derivative $f'$ in $[a,b]$, then $\omega(f;\delta) \leqq \delta\|f'\|_{C[a,b]}$.


## 1.3   Integral and averaging moduli

By $L_p[a,b]$ $(L_p[0,1] = L_p,\ L_1 = L)$, $1 \leqq p < \infty$ we denote the normed linear space of the integrable $(p)$ functions $f:[a,b] \to \mathbb{R}$ by the norm

$$\|f\|_p = \|f\|_{L_p[a,b]} = \left\{\frac{1}{b-a}\int_a^b |f(x)|^p\,\mathrm{d}x\right\}^{1/p}. \qquad (1.3.1)$$

We consider the function $f\in L_p[a,b]$ as something different from its equivalence class. This means that we consider the function $f$ as determined by its value $f(x)$ in each point $x\in[a,b]$.

The integral modulus ($L_p$-modulus, or $p$-modulus) of a function $f\in L_p[a,b]$ is the following function of $\delta\in[0,b-a]$

$$\omega(f;\delta)_{L_p} = \omega(f;\delta)_p = \sup\left\{\left[\frac{1}{b-a}\cdot\int_a^{b-h} |\Delta_h f(x)|^p\,\mathrm{d}x\right]^{1/p} : 0 \leqq h \leqq \delta\right\}$$

$$= \sup\left\{\left[\frac{1}{b-a}\cdot\int_a^{b-h} |f(x-h)-f(x)|^p\,\mathrm{d}x\right]^{1/p} : 0 \leqq h \leqq \delta\right\}. \qquad (1.3.2)$$

If the function $f:[a,b] \to \mathbb{R}$ is of bounded variation, then

$$\omega(f;\delta)_{L_1} = \omega(f;\delta)_1 \leqq \frac{\delta}{b-a} \bigvee_a^b f. \tag{1.3.3}$$

The local modulus of continuity of a function $f:[a,b] \to \mathbb{R}$ in a point $x \in [a,b]$ is defined as the function

$$\omega(f;x;\delta) = \sup\left\{ |f(t) - f(y)| : t, y \in \left[ x - \frac{\delta}{2}, x + \frac{\delta}{2} \right] \cap [a,b] \right\} \tag{1.3.4}$$

of the variable $\delta \in [0, b-a]$.

The averaged modulus of smoothness (or, $\tau_p$-modulus) of a bounded and measurable function $f:[a,b] \to \mathbb{R}$ (the set of all such functions is denoted by $M[a,b]$) is defined as the following function of the variable $\delta \in [0, b-a]$

$$\tau(f;\delta)_{L_p} = \tau(f;\delta)_p = \left\{ \frac{1}{b-a} \int_a^b \omega^p(f;x;\delta)\,dx \right\}^{1/p} \tag{1.3.5}$$

with $1 \leqq p < \infty$.

The $\tau$-modulus was introduced by Sendov [3] and Korovkin [4]. Dolženko and Sevast'anov introduced independently $\tau(f;\delta)_1 = \tau(f;\delta)$.

The $\tau_p$-moduli have the following properties (see [2]):

(i) $\tau(f;\delta)_p \leqq \tau(f;\delta')_p$, $0 < \delta \leq \delta' \leq b - a$ — monotonity;

(ii) $\tau(f+g;\delta)_p \leqq \tau(f;\delta)_p + \tau(g;\delta)_p$ — semi-additivity with respect to the function;

(iii) $\tau(f;\delta)_p \leqq \delta \| f' \|_p$ — estimation by means of the norm of the derivative;

(iv) $\tau(f;\delta)_1 = \tau(f;\delta) \leqq \dfrac{\delta}{b-a} \bigvee_a^b f$ — estimation by means of the variation of the function;

(v) $\tau(f;n\delta)_p \leqq n\tau(f;\delta)_p$ — homogeneity with respect to an integer multiplier $n \in \mathbb{N}$;

(vi) if $f \in M[a,b]$, then $\omega(f;\delta)_p \leqq \tau(f;\delta)_p \leqq \omega(f;\delta)$;

(vii) a necessary and sufficient condition for a function $f:[a,b] \to \mathbb{R}$ to be Riemann-integrable in the segment $[a,b]$ (the set of all such functions is denoted by $R[a,b]$) is

$$\lim_{\substack{\delta \to 0 \\ \delta > 0}} \tau(f;\delta) = 0. \tag{1.3.6}$$

## 1.4  Spaces and classes of functions

Along with other notation yet to be introduced, normed linear spaces $C[a,b]$ and $L_p[a,b]\,(1\leq p<\infty)$ of the continuous functions and of the functions with integrable $p$th power $f:[a,b]\to\mathbb{R}$, we introduce the normed linear space $L_\infty[a,b]\,(L_\infty[0,1]=L_\infty)$ of all measurable functions $f:[a,b]\to\mathbb{R}$, bounded up to a set with zero measure, with the essential supremum of $|f(x)|$ on the segment $[a,b]$

$$\|f\|_\infty = \|f\|_{L_\infty} = \operatorname{ess\,sup}\{|f(x)|:a\leqq x\leqq b\} = M \qquad (1.4.1)$$

as the norm, where $M$ is the smallest of the positive numbers $K$, for which the set $\{x:|f(x)|>K,\ x\in[a,b]\}$ has zero measure.

It is well known ([12], p 58) that if $f\in C$, then

$$\|f\|_C = \|f\|_\infty. \qquad (1.4.2)$$

Furthermore (see [1], p 21), if $f\in M[a,b]$, then

$$\lim_{p\to+\infty} \|f\|_p = \|f\|_\infty. \qquad (1.4.3)$$

Let $C^r(r\in\mathbb{N}_c)$ be the set of all $r$-times differentiable functions $f:[0,1]\to\mathbb{R}$, and $L_p^r(r\in\mathbb{N},\ 1\leq p<\infty)$ be the set of all functions $f:[0,1]\to\mathbb{R}$ having absolutely continuous $(r-1)$th derivative in $[0,1]$, $f^{(r)}\in L_p$.

The function classes $W_p^r(r\in\mathbb{N},\ 1\leq p<\infty)$ are determined by

$$W_p^r = \{f(x):f\in L_p^r,\ \|f^{(r)}\|_p \leqq 1\}.$$

Each modulus of continuity $\omega(\delta)$, $\delta\in[0,1]$, determines a subclass $H^\omega$ of the normed linear space $C$ consisting of all functions $f(x)$ for which the inequality

$$|f(x)-f(x')| \leqq \omega(|x-x'|) \qquad (1.4.4)$$

is fulfilled for each two points $x$ and $x'$ of $[0,1]$.

We denote by $W^r H^\omega(r\in\mathbb{N}_0,\ W^0H^\omega = H^\omega)$ the class of the $r$th indefinite integrals of the functions $f$ of $H^\omega$, i.e.

$$W^r H^\omega = \{f(x):f\in C^r,\ f^{(r)}\in H^\omega\}. \qquad (1.4.5)$$

In the case $\omega(t)=t^\alpha$, $0<\alpha\leqq 1$, $t\in[0,1]$, the classes $W^r H^\omega$ are denoted by $W^r H^\alpha$.

As is known ([12], p 59)

$$W_\infty^{r+1} = W^r H^1. \tag{1.4.6}$$

A set $\mathcal{M}$ of elements of the normed linear space $H(\mathcal{M} \subset H)$ is said to be convex, if

$$f \in \mathcal{M}, g \in \mathcal{M} \Rightarrow \alpha f + (1 - \alpha)g \in \mathcal{M} \ \forall \alpha \in [0, 1].$$

A set $\mathcal{M}$ is said to be centrally symmetric with the respect to the zero point, if

$$f \in \mathcal{M} \Rightarrow (-1)f \in \mathcal{M}.$$

For some of the later considerations the following assertion is of importance.

*Theorem A* (Korneichuk [6], pp 183–7, 24). Let $r(r \in \mathbb{N}_0)$ be fixed, and $\omega(t)$ be an arbitrary modulus of continuity. Then the class $W^r H^\omega$ is a convex, centrally symmetric set with respect to the zero in the normed linear space $C$.

The function class $Z_M \subset C[0, 1]$ with

$$\omega_2(f; \delta) \leqq M\delta, \qquad \delta \in [0, 1], \qquad M > 0 \tag{1.4.7}$$

is said to be ([7], p 142) a Zygmund class.

Let $D_2 = \{(x, y): 0 \leqq x \leqq 1, 0 \leqq y \leqq 1\}$ be the unit square in the two-dimensional Euclidean space $\mathbb{R}^2$. We denote by $C(D_2)$ the normed linear space of all continuous functions $f \cdot D_2 \to \mathbb{R}$ with the norm

$$\|f\|_{C(D_2)} = \sup\{|f(x, y)| : (x, y) \in D_2\}. \tag{1.4.8}$$

Let $H^{\omega_1, \omega_2}(D_2)$ be the function class of all functions $f \in C(D_2)$ such that for every two points $M = (x, y)$ and $M' = (x', y')$, from $D_2$ they satisfy the condition

$$|f(x, y) - f(x', y')| \leqq \omega_1(|x - x'|) + \omega_2(|y - y'|) \tag{1.4.9}$$

with some moduli of continuity $\omega_1(t)$ and $\omega_2(t)$.

Finally, let us introduce the class $H^\omega(D_2) \subset C(D_2)$ of the functions $f(x, y) = f(M)$, having the property that for every two points $M = (x, y)$ and $M' = (x', y')$, from $D_2$ they satisfy the condition

$$|f(M) - f(M')| \leqq \omega(\rho(M, M')) \tag{1.4.10}$$

with a given modulus of continuity $\omega(\delta)$, where $\rho(M, M')$ denotes the Euclidean distance between the points $M$ and $M'$.

## 1.5   Statement of the extremal problem

Let $\mathcal{M}$ be a given class of functions of the normed linear space $H$, $\mathcal{M} \subset H$, $m$ be a fixed number. We assume that $L$, $L_1$, $L_2$, ..., $L_m$ are $m+1$ given functionals defined on $\mathcal{M}$ with values in $\mathbb{R}$.

In many problems, of importance for applications one has to calculate approximately the value $Lf$ of a functional $L$ for a fixed function $f \in \mathcal{M}$ provided one has at one's disposal only the elements of the matrix

$$T = (L_1 f, L_2 f, \ldots, L_m f) \tag{1.5.1}$$

(the information for the function $f$). It is assumed that the information $T$ for each function $f \in \mathcal{M}$ is either known, or can be easily calculated.

In this case we say that the functional $L$ is to be restored in the class $\mathcal{M}$, based on the information $T$.

By $S$ we denote the set of all possible methods for a restoration of the functional $L$ in the class $\mathcal{M}$, using the information $T$ only. The error of each such method $\Phi \in S$ is determined by

$$R(\mathcal{M}, T, \Phi) = \sup\{|Lf - \Phi f| : f \in \mathcal{M}\}.$$

From the point of view of the applications, of interest are those methods $\Phi \in S$ for restoration of the functional $L$ in $\mathcal{M}$ on $T$ which have the smallest error.

Let us denote

$$R(\mathcal{M}, T) = \inf\{R(\mathcal{M}, T, \Phi) : \Phi \in S\} = \inf_{\Phi \in S} \sup_{f \in \mathcal{M}} |Lf - \Phi f|. \tag{1.5.2}$$

Each method $\Phi^* \in S$, which is a solution of the extremal problem (1.5.2), i.e. for which the equality

$$R(\mathcal{M}, T) = R(\mathcal{M}, T, \Phi^*) = \sup_{f \in \mathcal{M}} |Lf - \Phi^*| \tag{1.5.3}$$

holds is said to be an optimal method for restoration of the functional $L$ in the function class $\mathcal{M}$ on the information $T$ or shortly, a best method for $L$ in $\mathcal{M}$ on $T$.

We will consider the extremal problem (1.5.2) for the linear functionals

$$Lf = \int_0^1 f(x)\,\mathrm{d}x; \tag{1.5.4}$$

$$Lf = f(x), \qquad x \in [0, 1]; \tag{1.5.5}$$

$$Lf = f'(x), \qquad x \in [0, 1] \tag{1.5.6}$$

in the function class $\mathcal{M} = W^r H^\omega$ for an arbitrary modulus of continuity $\omega(t)$ and an arbitrary fixed non-negative integer $r$, provided $L_1 f, L_2 f, \ldots, L_m f \, (m = n(r+1))$ are the values of the function $f \in W^r H^\omega$ and of its derivative up to the $r$th order in the nodes of a given mesh $X = (x_1, x_2, \ldots, x_n)$, i.e. on the information $T_r(f, X)$ determined in (1.1.4).

For such functionals a basic tool in treating an extremal problem (1.5.2) is the following lemma of Smolyak [8] (see also [9], or [10], p 107).

*Lemma* (Smolyak). Let the linear functionals $L, L_1, L_2, \ldots, L_m (m \in \mathbb{N})$ be defined in the linear space $H$. Let $\mathcal{M}$ be a convex, centrally symmetric body in $H$ with the zero as the centre of symmetry, and

$$\sup\{Lf : f \in \mathcal{M}_0\} < \infty \tag{1.5.7}$$

where

$$\mathcal{M}_0 = \{f(x) : f \in \mathcal{M}, L_k f = 0, k = 1, 2, \ldots, m\} = \{f(x) : f \in \mathcal{M}, T = 0\}. \tag{1.5.8}$$

Then, there exist constants $C_1, C_2, \ldots, C_m$ such that the equality

$$R(\mathcal{M}, T) = \sup\left\{ \left| Lf - \sum_{i=1}^{m} C_i L_i f \right| : f \in \mathcal{M} \right\} \tag{1.5.8}$$

holds, i.e. there exists a linear method which is the best.

**Corollary 1.5.1** *Under the hypothesis and notations in Smolyak's lemma, the following holds*

$$R(\mathcal{M}, T) = \sup\{Lf : f \in \mathcal{M}_0\}. \tag{1.5.10}$$

This useful assertion follows from the duality relations of Nikol'skii [11] also (in addition see [12], p 114, Theorem 3.3.6).

In the case of functionals (1.5.6) and (1.5.6) the error of the best method for restoration of $L$ in $\mathcal{M}$ on $T$ depends on the choice of the fixed point $x \in [0, 1]$, i.e.

$$R(\mathcal{M}, T) = R(\mathcal{M}, T, x), \qquad x \in [0, 1]. \tag{1.5.11}$$

Letting $x$ vary over the segment $[0, 1]$, we obtain various errors. Then it is natural to look for an estimate of them.

Let us put

$$R(\mathcal{M}, T)_C = \|R(\mathcal{M}, T; \cdot)\|_C = \sup\{R(\mathcal{M}, T; x) : 0 \leqq x \leqq 1\} \tag{1.5.12}$$

and

$$R(\mathcal{M}, T)_p = \| R(\mathcal{M}, T; \cdot) \|_p = \left\{ \int_0^1 [R(\mathcal{M}, T; x)]^p \, dx \right\}^{1/p}, \qquad p > 0 \quad (1.5.13)$$

under the natural assumption that the $C$, or $L_p$ norm exists (in the case $0 < p < 1$ instead of norm, we speak of a quasi-norm). The numbers $R(\mathcal{M}, T)_C$ and $R(\mathcal{M}, T)_p$ are called the error of the best method for $L$ in $\mathcal{M}$ on $T$ in $C$ or $L_p(p > 0)$ norm, respectively.

In many cases, including the cases treated in this book, the information $T$ is composed of values of the function $f \in \mathcal{M}$ and of its derivatives up to a given order at nodes $x_i (i = 1, 2, \ldots, n; n \in \mathbb{N})$ of a given mesh $X = (x_1, x_2, \ldots, x_n)$ of the segment $[0, 1]$. In such cases the information $T$ changes with the varying of the mesh $X$, i.e. $T = T(f; X)$, and the error of the corresponding best method also changes with the mesh:

$$R(\mathcal{M}, T) = R(\mathcal{M}, T(f; X)) = : R(\mathcal{M}, X),$$

$$R(\mathcal{M}, T)_C = R(\mathcal{M}, T(f; X))_C = : R(\mathcal{M}, X)_C, \qquad (1.5.14)$$

$$R(\mathcal{M}, T)_p = R(\mathcal{M}, T(f; X))_p = : R(\mathcal{M}, X)_p, \qquad p > 0.$$

It is natural to state the extremal problem for a search of such a mesh $X$ for which the corresponding error of the best method for $L$ in $\mathcal{M}$ on $T(f; X)$ is the smallest possible.

Let $(X)_v$ be a given set of meshes $X$ over the segment $[0, 1]$. Then, let us denote

$$R_n^v(\mathcal{M}) = \inf\{R(\mathcal{M}, X): X \in (X)_v\},$$

$$R_n^v(\mathcal{M})_C = \inf\{R(\mathcal{M}, X)_C: X \in (X)_v\}, \qquad (1.5.15)$$

$$R_n^v(\mathcal{M})_p = \inf\{R(\mathcal{M}, X)_p: X \in (X)_v\}.$$

If there exists a mesh $X_v \in (X)_v$ for which the equality

$$R_n^v(\mathcal{M}) = R(\mathcal{M}, X_v),$$

$$\text{or} \qquad R_n^v(\mathcal{M})_C = R(\mathcal{M}, X_v)_C, \qquad (1.5.16)$$

$$\text{or} \qquad R_n^v(\mathcal{M})_p = R(\mathcal{M}, X_v)_p, \qquad p > 0$$

holds, it is said to be an optimal mesh for $L$ in $\mathcal{M}$ on $T(f; X)$ for the corresponding case or metric. If $\Phi(T(f;X))$ is a best method for $L$ in $\mathcal{M}$ on $T(f;X)$, then the method $\Phi(T(f;X_\nu))$ relative to an optimal mesh $X_\nu$ is said to be an optimal method for $L$ in $\mathcal{M}$ on $T(f;X)$ (in the corresponding case or metric).

The best (optimal) method for restoration of the functional (1.5.4) in $\mathcal{M}$ on $T(f;X)$ is said to be the best (optimal) quadrature formula for the integral (4) in $\mathcal{M}$ on $T(f;X)$ also.

If for a function $f^*\in\mathcal{M}$ the equality

$$R_n^\nu(\mathcal{M}) = |Lf^* - \Phi^*(T(f^*;X_\nu)|,$$

$$\text{or} \qquad R_n^\nu(\mathcal{M})_C = \|Lf^* - \Phi^*(T(f^*;X_\nu);\cdot)\|_C, \qquad\qquad (1.5.17)$$

$$\text{or} \qquad R_n^\nu(\mathcal{M})_p = \|Lf^* - \Phi^*(T(f^*;X_\nu);\cdot)\|_p, \qquad p > 0$$

holds, then it is said to be an extremal function for the corresponding case.

The extremal problem (1.5.2), (1.5.15) stems from the classical works of Favard [13], Kolmogorov [14] and Nikol'skii [11].

In this book we consider the extremal problem (1.5.2), (1.5.15) in the normed linear space of continuous functions of one, or several variables for functionals (1.5.4)–(1.5.6) in function classes, determined by an arbitrary modulus of continuity and a corresponding information.

There are many works devoted to an optimal restoration of functionals in the classes $W_p^r(1 \leq p \leq \infty, r\in\mathbb{N}_0)$. In the first place should be mentioned the investigations of Sard [15], [16], Nikol'skii and his pupils (see [1] and the Appendix to this book). In this direction see also the investigations of Boyanov [17].

Whilst the investigations on the optimal recovery of functionals in the classes $W_p^r$ are of more or less definite form, there are only isolated papers in classes $W^r H^\omega(r\in\mathbb{N}_0, \omega(t)$ is an arbitrary modulus of continuity). Let us survey them in some detail.

In 1951 Turezkii [18] had proved the following.

*Theorem B* (Turezkii [18]). Let $n\in\mathbb{M}$ and $0 < \alpha \leq 1$. Out of all quadrature formulas of the form

$$\int_0^1 f(x)\mathrm{d}x \approx \sum_{i=1}^n C_i f(x_i)$$

using the information

$$T_0(f;X) = (f(x_1), f(x_2), \ldots, f(x_n))$$

only, the optimal one in the function class $H^\alpha$, $0 < \alpha \leq 1$ is the quadrature formula of the middle rectangles

$$\Phi(T_0(f; X_0)) = \frac{1}{n} \sum_{i=1}^{n} f\left(\frac{2i-1}{2n}\right) \qquad (1.5.18)$$

for which

$$R_n^0(H^\alpha) = \frac{1}{(\alpha + 1)(2n)^\alpha} \qquad (1.5.19)$$

In 1968 Korneichuk [19] (see also [1], p 206) showed that the quadrature formula of the middle rectangles (1.5.18) is optimal in the class $H^\omega$, and also for arbitrary modulus of continuity. For the error the explicit expression

$$R_n^0(H^\omega) = \int_0^1 \omega\left(\frac{u}{2n}\right) du \qquad (1.5.20)$$

is obtained (for $\omega(t) = t^\alpha$, $0 < \alpha \leq 1$ which implies (19)). In the same work Korneichuk also obtained a multi-dimensional analogue of formula (1.5.18) for a lattice of nodes.

As for the optimal recovery of functions of the classes $W^r H^\omega$ based on information $T_r(f; X)$, a result of Korneichuk for the class $H^\omega$ where the modulus of continuity $\omega(t)$ is convex from above, is known in the $L_p$-metric for $p \in [1, 3]$ (see, e.g. [12], p 295, Theorem 6.4.3, the part concerning the non-periodic case).

Among with the general setting of the extremal problem already considered, we will now also consider a restricted version of it.

With each mesh $X = (x_1, x_2, \ldots, x_n)$, $0 \leq x_1 < x_2 < \ldots < x_n \leq 1$, we associate a second mesh $B = (b_0, b_1, \ldots, b_{n-1}, b_n)$ uniquely determined from $X$ according to the formulas

$$b_0 = 0, \qquad b_n = 1, \qquad b_i = (x_i + x_{i+1})/2, \qquad i = 1, 2, \ldots, n-1 \quad (1.5.21)$$

Let $\chi_i(x)(i = 1, 2, \ldots, n)$ be the indicator function of the interval $[b_{i-1}, b_i)$† $(i = 1, 2, \ldots, n)$, i.e.

$$\chi_i(x) = \begin{cases} 1 & \text{for } x \in [b_{i-1}, b_i), \\ 0 & \text{for } x \notin [b_{i-1}, b_i). \end{cases} \qquad (1.5.22)$$

The functions

$$\chi_1(x), \chi_2(x), \ldots, \chi_n(x) \qquad (1.5.23)$$

† By definition, $[\alpha, \beta) = [\alpha, \beta]$ for $\beta = 1$ and $\alpha \in [0, 1]$.

are non-negative and they form a partition of the unit in the segment $[0, 1]$:

$$\chi_i(x) \geqq 0 \quad (i = 1, 2, \ldots, n); \qquad \sum_{i=1}^{n} \chi_i(x) = 1, \qquad x \in [0, 1] \quad (1.5.24)$$

They form a system of fundamental quasi-splines. The quasi-spline of $r$th degree generated by them we denote as

$$\Phi^r(x) = \Phi^r(f, X; x) = \sum_{i=1}^{n} \sum_{k=0}^{n} \chi_i(x) \frac{f^{(k)}(x_i)}{k!} (x - x_i)^k, \qquad x \in [0, 1]. \quad (1.5.25)$$

We denote the following

$$\Delta(W^r H^\omega, X; x) = \sup\{|f(x) - \Phi^r(f, X; x)|: f \in W^r H^\omega\}; \quad (1.5.26)$$

$$\Delta(W^r H^\omega, X)_C = \sup\{\Delta(W^r H^\omega, X; x): 0 \leqq x \leqq 1\}; \quad (1.5.27)$$

$$\Delta(W^r H^\omega, X)_p = \left\{\int_0^1 [\Delta(W^r H^\omega. X; x)]^p \, dx\right\}^{1/p}, \qquad p > 0; \quad (1.5.28)$$

$$\Delta_n^v(W^r H^\omega)_C = \inf\{\Delta(W^r H^\omega, X)_C: X \in (X)_v\}; \quad (1.5.29)$$

$$\Delta_n^v(W^r H^\omega)_p = \inf\{\Delta(W^r H^\omega, X)_p: X \in (X)_v\}. \quad (1.5.30)$$

If for a mesh $X_v \in (X)_v$ we have the equality

$$\Delta_n^v(W^r H^\omega)_C = \Delta(W^r H^\omega, X_v)_C \quad (1.5.31)$$

or

$$\Delta_n^v(W^r H^\omega)_p = \Delta(W^r H^\omega, X_v)_p, \qquad p > 0 \quad (1.5.32)$$

then we say that it is optimal in the recovery of the function $f$ in the class $W^r H^\omega$ using the method (1.5.25), and the quasi-spline $\Phi^r(f, X_v; x)$ constructed on formula (1.5.25) for this mesh is said to be an optimal method of the form (1.5.25) in $C$ or $L_p$ $(p > 0)$ metrics.

## 1.6  Meshes and their characteristics

The set of all meshes $X = (x_1, x_2, \ldots, x_n)$ satisfying the condition (1.1.1) will be denoted by $(X)_0$; further, we write $X \in (X)_v, v = 1, 2, 3$ when $X \in (X)_0$ and

$$x_1 = 0 \qquad \text{for} \quad v = 1 \quad (1.6.1)$$

or

$$x_1 = 0, \qquad x_n = 1, \qquad n > 1 \qquad \text{for} \quad v = 2 \tag{1.6.2}$$

or

$$x_n = 1 \qquad \text{for} \quad v = 3 \tag{1.6.3}$$

Let $A = (a_0, a_1, a_2, \ldots, a_n)$ be a second mesh of nodes in the segment $[0, 1]$ subjected to the restriction

$$0 = a_0 < a_1 < a_2 < \ldots < a_n = 1. \tag{1.6.4}$$

Then we say that the pair of meshes $(A, X)$ belongs to the set $(A, X)_v$ and write $(A, X) \in (A, X)_v$, provided that along with conditions (1.1.1) and (1.6.4) the relation $X \in (X)_v$, $v = 0, 1, 2, 3$, is true.

Let us introduce the mesh pairs $(A_v, X_v) \in (A, X)_v$, $A_v = (a_i^{(v)})$, $X_v = (x_i^{(v)})$ by means of the equations

$$A_0 = \left( 0, \frac{1}{n}, \frac{2}{n}, \ldots, 1 \right), \qquad a_i^{(0)} = \frac{i}{n} \quad (i = 0, 1, 2, \ldots, n - 1)$$

$$X_0 = \left( \frac{1}{2n}, \frac{3}{2n}, \ldots, \frac{2n-1}{2n} \right), \qquad x_i^{(0)} = \frac{2i-1}{2n} \quad (i = 1, 2, \ldots, n); \tag{1.6.5$_0$}$$

$$A_1 = \left( 0, \frac{1}{2n-1}, \ldots, 1 \right), \qquad a_i^{(1)} = \frac{2i-1}{2n-1} \quad (i = 1, 2, \ldots, n - 1)$$

$$X_1 = \left( 0, \frac{2}{2n-1}, \ldots, \frac{2n-2}{2n-1} \right), \qquad x_i^{(1)} = \frac{2i-2}{2n-1} \quad (i = 1, 2, \ldots, n); \tag{1.6.5$_1$}$$

$$A_2 = \left( 0, \frac{1}{2n-2}, \ldots, 1 \right), \qquad a_i^{(2)} = \frac{2i-1}{2n-2} \quad (i = 1, 2, \ldots, n - 1)$$

$$X_2 = \left( 0, \frac{1}{n-1}, \ldots, 1 \right), \qquad x_i^{(2)} = \frac{i-1}{n-1} \quad (i = 1, 2, \ldots, n); \tag{1.6.5$_2$}$$

$$A_3 = \left( 0, \frac{2}{2n-1}, \ldots, 1 \right), \qquad a_i^{(3)} = \frac{2^i}{2n-1} \quad (i = 1, 2, \ldots, n - 1)$$

$$X_3 = \left( \frac{1}{2n-1}, \frac{3}{2n-1}, \ldots, 1 \right), \qquad x_i^{(3)} = \frac{2i-1}{2n-1} \quad (i = 1, 2, \ldots, n). \tag{1.6.5$_3$}$$

Note that $X_0$ determined in $(1.6.5_1)$ is the classical optimal mesh (see, e.g. [21], p 62, Theorem 2).

Characteristic for the double meshes are the so-called deviations. The corresponding definitions follow.

Let $\chi_\gamma(x)$ be the characteristic function of the interval $[0, \gamma)$, $\gamma \in [0, 1]$. (Here we use the convention $[\alpha, \beta) = [\alpha, \beta]$ if $\beta = 1$ and $\alpha \in [0, 1]$ also.)

We define the local deviation $R(A, X; \gamma)$ for the mesh pairs $(A, X) \in (A, X)_0$ in the point $\gamma \in [0, 1]$ by

$$R(A, X; \gamma) = F_n(\gamma) - \gamma \qquad (1.6.6)$$

with

$$F_n(\gamma) = \sum_{i=1}^{n} (a_i - a_{i-1}) \chi_\gamma(x_i), \qquad \gamma \in [0, 1]. \qquad (1.6.7)$$

$(A, X) \in (A, X)_0$ and from (1.6.7) it follows that the function $F_n(\gamma)$, $\gamma \in [0, 1]$ is an increasing stepwise function with discontinuity points $x_1, x_2, \ldots, x_n$ and

$$F_n(\gamma) = \begin{cases} a_0 & \text{for} \quad \gamma \in [x_0, x_1] \\ a_i & \text{for} \quad \gamma \in [x_i, x_{i+1}] \quad (i = 1, 2, \ldots, n) \end{cases} \qquad (1.6.8)$$

where for the sake of convenience the nodes $x_0 = 0$ and $x_{n+1} = 1$ are introduced.

Following Proinov [20] the notion of the uniform (or, $L_\infty$) deviation of the mesh pair $(A, X) \in (A, X)_0$ is defined by the formula

$$D(A, X)_\infty = \sup\{|R(A, X; \gamma)| : 0 \le \gamma \le 1\}. \qquad (1.6.9)$$

The integral (or, $L_p$, $p > 0$) deviation of the same mesh pair is defined by the formula

$$D(A, X)_p = \left[ \int_0^1 |R(A, X; x)|^p \, dx \right]^{1/p} = \left[ \int_0^1 |F_n(\gamma) - \gamma|^p \, d\gamma \right]^{1/p}, \quad p > 0. \qquad (1.6.10)$$

From (1.6.6), (1.6.9) and (1.6.10), using the well-known equality

$$\lim_{p \to \infty} \left[ \int_0^1 |F_n(x) - x|^p \, dx \right]^{1/p} = \sup\{|F_n(x) - x| : 0 \le x \le 1\} \qquad (1.6.11)$$

(see, e.g. [21], p 60, or (1.4.3)), we obtain the following.

**Corollary 1.6.1**   *For each mesh pair $(A, X) \in (A, X)_0$ the equality*

$$\lim_{p \to \infty} D(A, X)_p = D(A, X)_\infty \tag{1.6.12}$$

*holds.*

Note that the uniform deviation $D(A_0, X)_\infty$ was introduced by Van der Korput [22], and the $L_2$ deviation $T = D(A_0, X)_2$ by Roth [23].

Proinov [20] has shown that

$$D(A, X)_\infty = \max\{|a_i - x_i|, |a_{i-1} - x_i| : 1 \leqq i \leqq n\}. \tag{1.6.13}$$

From this equality it follows that if $x \in [a_{i-1}, a_i]$, $i = 1, 2, \ldots, n$, then

$$|x - x_i| \leqq D(A, X)_\infty. \tag{1.6.14}$$

## 1.7   Survey of the results

In Chapter 2 methods for recovery of functions of the classes $W^r H^\omega$ on given information $T_r(f; X)$ are considered. The considerations are as in uniform and integral metrics. These methods are optimal in some sense and contain some of the extremal problems; (1.5.15) or (1.5.30).

In Section 2.1 the following assertion, used in further considerations, is proved.

*Theorem 2.1.1* ([24], [28]). If the function $\varphi : [0, 1] \to \mathbb{R}$ is (strictly) increasing, then for each mesh pair $(A, X) \in (A, X)_v$ the inequality

$$L_n(\varphi; A, X) \geqq L_n(\varphi; A_v, X_v) \tag{1.7.1}$$

where

$$L_n(\varphi; A, X) \stackrel{\text{def}}{=} \sum_{i=1}^{n} \int_{a_{i-1}}^{a_i} \varphi(|x - x_i|) \mathrm{d}x, \qquad n \in \mathbb{N} \tag{1.7.2}$$

holds. The equality in (1.7.1) is attained (only) for the mesh pair $(A_v, X_v)$, determined in $(1.6.5_v)$ ($v = 0, 1, 2, 3$) and

$$L_n(\varphi; A_v, X_v) = \int_0^1 \varphi(u \cdot D(A_v, X_v)_\infty) \mathrm{d}u \tag{1.7.3}$$

where $D(A, X)_\infty$ is the uniform deviation of the mesh pair $(A, X)$.

As a first application of this theorem we obtain lower bounds for the deviation of mesh pairs.

*Corollary 2.1.3* ([24]). For each mesh pair $(A, X) \in (A, X)_v$ the inequality

$$D(A, X)_p \geqq D(A_v, X_v)_p, \qquad 0 < p \leqq \infty \tag{1.7.3'}$$

holds with the equality sign only for the mesh pair $(A_v, X_v)$ $(v = 0, 1, 2, 3)$, defined in $(1.6.5_v)$, or, which is the same, for $0 < p \leqq \infty$ and $v = 0, 1, 2, 3$.

In the special case $v = 0$ and $1 \leqq p < \infty$, from Corollary 2.1.3 we obtain the following assertion (see [21], p 62, Theorem 2). For each mesh pair $(A, X) \in (A, X)_0$ the inequality

$$D(A, X)_p \geqq \frac{1}{2n\sqrt[p]{1 + p}}, \qquad 1 \leqq p < \infty$$

with equality only for the mesh pair $(A_0, X_0)$, determined in $(1.6.5_0)$ holds. For $p = 2$ the estimation of the $L_2$ deviation (see [23], or, e.g. [58]: For each mesh $X \in (X)_0$ the inequality

$$T(X) = D(A_0, X)_2 \geqq \frac{1}{n\sqrt{12}}$$

with the equality sign only for the mesh $X_0$, determined in $(1.6.5_0)$ holds.

Again, from Corollary 2.1.3 for $p = \infty$ we obtain a known result. For each mesh pair $(A, X) \in (A, X)_0$ the inequality

$$D(A, X)_\infty \geqq 1/2n$$

with equality only for $(A_0, X_0)$ holds (see, e.g. [21], p 64, Theorem 2').

In Section 2.2 a standard function for the class $H^\omega$ (with an arbitrary modulus of continuity $\omega(t)$) and for the mesh $X = (x_1, x_2, \ldots, x_n)$ is introduced

$$f_0(\omega, X; x) = \sum_{i=1}^{n} \chi_i(x) \cdot \omega(|x - x_i|) \tag{1.7.4}$$

where $\chi_i(x)$ is the characteristic function of the interval $[b_{i-1}, b_i)$, and the mesh $B = (b_0, b_1, \ldots, b_n)$ is uniquely determined by the mesh $X$ according to the formulas

$$b_0 = 0, \qquad b_n = 1, \qquad b_i = \frac{x_i + x_{i+1}}{2} \quad (i = 1, 2, \ldots, n - 1) \tag{1.7.5}$$

with the convention $[b_{n-1}, b_n) = [b_{n-1}, b_n]$.

In Lemma 2.2.1 the properties of this function, used in the following considerations, are studied.

In Section 2.3 a best method for recovery of a function of the class $H^\omega$ on information $T_0(f;X)$ is considered.

*Theorem 2.3.1* ([25]). Let $n\in\mathbb{N}$, $X\in(X)_0$, $\omega(t)$ be an arbitrary modulus of continuity, and $x\in[0,1]$ be an arbitrary fixed point.

Out of all possible methods for recovery of a functional $Lf = f(x)$ in the class $H^\omega$ on information $T_0(f,X)$, $X\in(X)_0$ the best is

$$\Phi^0(x) \equiv \Phi^0(f, X; x) = \sum_{i=1}^{n} \chi_i(x)\cdot f(x_i) \qquad (1.7.6)$$

where $\chi_i(x)$ denotes the same characteristic function as in (1.7.4).

For each function $f\in H^\omega$ and for every $x\in[0,1]$ holds the inequality

$$|f(x) - \Phi^0(f, X; x)| \leqq f_0(\omega, X; x) \qquad (1.7.7)$$

with equality for the standard function (1.7.4).

For the error of the best method the equality

$$R(H^\omega, X; x) = \sup_{f\in H^\omega} |f(x) - \Phi^0(f, X; x)| = f_0(\omega; X; x). \qquad (1.7.8)$$

holds.

*Corollary 2.3.1* ([25]). Let $n\in\mathbb{N}$, $X\in(X)_0$, and $\omega(t)$ be an arbitrary modulus of continuity. Then the error of the best method for recovery of functions of the class $H^\omega$ on the information $T_0(f,X)$, $X\in(X)_0$ in uniform and in integral $(L_p, p>0)$ metrics is given by

$$R(H^\omega, X)_C = \|R(H^\omega, X, \cdot)\|_C = \omega(D(A, X)_\infty) \qquad (1.7.9)$$

and

$$R(H^\omega, X)_{L_p} = \|R(H^\omega, X; \cdot)\|_{L_p} = [L_n(\omega^p(\cdot), B, X)]^{1/p}, \quad p>0 \qquad (1.7.10)$$

correspondingly, where the mesh $B$ is determined in (1.7.5), $D(A, X)_\infty$ is the uniform deviation of the mesh pair $(A, X)$, and $L_n(\varphi; A, X)$ is defined in (1.7.2).

In Section 2.4 optimal methods for recovery of functions of the class $H^\omega$ on information $T_0(f;X)$ are considered.

*Theorem 2.4.1* ([25]). Let $n\in\mathbb{N}$, $v\in\{0,1,2,3\}$, $X\in(X)_v$ and $\omega(t)$ be an arbitrary modulus of continuity. Out of all methods for recovery of functions of the class $H^\omega$ on information $T_0(f;X)$, $X\in(X)_v$ an optimal one for uniform and integral metrics is the method $\Phi^0(f, X_v; x)$ constructed in (1.7.6) for the set $X = X_v$, determined in $(1.6.5_v)$.

Furthermore

$$R_n^{v}(H^{\omega})_C = \inf_{X \in (X)_v} R(H^{\omega}, X)_C = \omega(D(A_v, X_v)_\infty) \qquad (1.7.11)$$

and

$$R_n^{v}(H^{\omega})_{L_p} = \inf_{X \in (X)_v} R(H^{\omega}, X)_{L_p} = \left[\int_0^1 \omega^p(u.D(A_v, X_v)_\infty)du\right]^{1/p}, \quad p > 0. \qquad (1.7.12)$$

Extremal is the standard function $f_0(\omega, X_v; x)$.

The proof of this theorem is based both on the optimization theorem 2.1.1 and Corollary and of the following lemma, which has it's own importance.

*Lemma 2.4.1.* Let $P$ be an arbitrary non-empty set, $\mathbb{R}_+ = \{x : x \in \mathbb{R}, x \geq 0\}$, $\varphi : P \to \mathbb{R}_+$ and $\omega(t)$ be an arbitrary modulus of continuity. Then the equality

$$\inf\{\omega(\varphi(x)): x \in P\} = \omega(\inf\{\varphi(x): x \in P\}) \qquad (1.7.13)$$

holds.

Let us write down explicitly the assertion of Theorem 2.4.1 for various meshes:

(*a*) for meshes $X = (x_1, x_2, \ldots, x_n)$ without additional restrictions (i.e. for $X = (X)_0$):

$$R_n^{0}(H^{\omega})_{L_p} = \sup_{f \in H^{\omega}} \|f - \Phi^0(f, X_0; \cdot)\|_{L_p}$$

$$= \left\{\int_0^1 \left[\omega\left(\frac{u}{2n}\right)\right]^p du\right\}^{1/p}, \qquad p > 0; \qquad (1.7.14)$$

(*b*) for meshes $X = (x_1, x_2, \ldots, x_n)$ with the restriction $x_1 = 0$ (i.e. $X \in (X)_1$):

$$R_n^{1}(H^{\omega})_{L_p} = \sup_{f \in H^{\omega}} \|f - \Phi^0(f, X_1; \cdot)\|_{L_p}$$

$$= \left\{\int_0^1 \left[\omega\left(\frac{u}{2n-1}\right)\right]^p du\right\}^{1/p}, \qquad p > 0; \qquad (1.7.15)$$

(*c*) for meshes $X = (x_1, x_2, \ldots, x_n)$ with the restrictions $x_1 = 0$, $x_n = 1$ and

$n > 1$ (i.e. $X \in (X)_2$):

$$R_n^2(H^\omega)_{L_p} = \sup_{f \in H^\omega} \| f - \Phi^0(f, X_2; \cdot) \|_{L_p}$$

$$= \left\{ \int_0^1 \left[ \omega\left( \frac{u}{2n-2} \right) \right]^p du \right\}^{1/p}, \qquad p > 0; \quad (1.7.16)$$

(*d*) for meshes $X = (x_1, x_2, \ldots, x_n)$ with the restriction $x_n = 1$ (i.e. $X \in (X)_3$):

$$R_n^3(H^\omega)_{L_p} = \sup_{f \in H^\omega} \| f - \Phi^0(f, X_3; \cdot) \|_{L_p}$$

$$= \left\{ \int_0^1 \left[ \omega\left( \frac{u}{2n-1} \right) \right]^p du \right\}^{1/p}, \qquad p > 0. \quad (1.7.17)$$

In the special case $p = \infty$ for the same meshes in the uniform metrics we have:

$$R_n^0(H^\omega)_C = \sup_{f \in H^\omega} \| f(\cdot) - \Phi^0(f, X_0; \cdot) \|_C = \omega\left( \frac{1}{2n} \right); \qquad (1.7.18)$$

$$R_n^1(H^\omega)_C = \sup_{f \in H^\omega} \| f(\cdot) - \Phi^0(f, X_1; \cdot) \|_C = \omega\left( \frac{1}{2n-1} \right); \qquad (1.7.19)$$

$$R_n^2(H^\omega)_C = \sup_{f \in H^\omega} \| f(\cdot) - \Phi^0(f, X_2; \cdot) \|_C = \omega\left( \frac{1}{2n-2} \right); \qquad (1.7.20)$$

$$R_n^3(H^\omega)_C = \sup_{f \in H^\omega} \| f(\cdot) - \Phi^0(f, X_3; \cdot) \|_C = \omega\left( \frac{1}{2n-1} \right). \qquad (1.7.21)$$

For a comparison we reproduce results of Malozemov [59], [69], Storchai [60] and Loginov [61] for approximation of functions with various interpolation splines using the same information.

A function $S_1(x)$ is said to be a spline of the first order on the mesh†

$$\Delta : 0 = x_0 < x_1 < \ldots < x_n = 1$$

when it is continuous on $[0, 1]$ and it coincides with a linear function on

† The mesh $\Delta$ has a node more than the mesh itself.

each of the segments $[x_{i-1}, x_i]$ $(i = 1, 2, \ldots, n)$. Given in the following three theorems, denoted by $S_1(x)$, is the spline of the first order on the mesh $(0, 1/n, 2/n, \ldots, 1)$ interpolating the function $f \in C[0,1]$ in the points $x_k = k/n$ $(k = 0, 1, \ldots, n)$.

*Theorem* (Malozemov [59], [69]). Let $\omega(t)$ be a convex modulus of continuity. Then

$$\sup_{f \in H^\omega} \|f - S_1\|_C = \omega\left(\frac{1}{2n}\right); \tag{1.7.22}$$

$$\sup_{f \in WH^\omega} \|f - S_1\|_C = \frac{1}{4} \int_0^{1/n} \omega(t)dt. \tag{1.7.23}$$

A generalization of the equality (1.7.22) is given by Storchai [60].

*Theorem* (Storchai [60]). Let $\omega(t)$ be a convex modulus of continuity. Then

$$\sup_{f \in H^\omega} \|f - S_1\|_{L_p} = \left\{ \int_0^1 \left[\omega\left(\frac{u}{2n}\right)\right]^p du \right\}^{1/p}, \qquad 1 \leq p < \infty. \tag{1.7.24}$$

The equality (1.7.22) follows from (1.7.24) letting $p \to \infty$.

*Approximations of functions of the class $H^\omega$ for an arbitrary modulus of continuity $\omega(t)$ by splines of the first degree are considered in the work* [61] of Loginov (by Hermitian splines—Velikin [40], by $\varphi$-splines—Storchai and Ligun [39], etc.).

*Theorem* (Loginov [61]). Let $\omega(t) \not\equiv 0$ be an arbitrary modulus of continuity. Then

$$I(\omega, n) = \sup_{f \in H\omega} \frac{\|f - S_1\|_C}{\omega(1/2n)} < 3/2 \tag{1.7.25}$$

and

$$\sup_{\omega(t) \not\equiv 0} I(\omega, n) = 3/2. \tag{1.7.26}$$

Now, we are in a position to compare (1.7.20) with (1.7.22) and (1.7.25) with (1.7.26) as follows:
(i) they use the same information;
(ii) equation (1.7.20) is valid for an arbitrary modulus of continuity, while (1.7.20) is valid for a convex modulus only;
(iii) estimate (1.7.20) is better that eatimate (1.7.25);
(iv) estimate (1.7.20) has the optimality property.

Estimate (1.7.20) has the same advantages with respect to the Hermitian splines [40], $\varphi$-splines [39], quasi-splines, etc.

As for estimate (1.7.16), one should note that:

(i) it uses the same information as (1.7.24);

(ii) equation (1.7.16) is valid for an arbitrary modulus of continuity, while (1.7.24) is valid for a convex one only;

(iii) estimate (1.7.16) has the optimality property;

(iv) there exists an analogue of (1.7.16) for $p = 1$ for $\varphi$-splines (see [39] and [40]) as follows.

*Theorem* (Storchai and Ligun [39]). For an arbitrary modulus of continuity, $\omega(t)$, the equation†

$$\sup_{f \in H^{\omega}} \| f - \rho_n(f; \cdot) \|_L = \int_0^1 \omega(t/2n)\mathrm{d}t \tag{1.7.27}$$

is true where $\rho_n(f; x)$ is a $\varphi$-spline of Storchai and Ligun [39] for the uniform mesh $x_k = k/n$ $(k = 0, 1, \ldots, n)$.

In Section 2.5 methods of a special fixed form, viz. (1.5.25) for recovery of functions of the classes $W^r H^{\omega}$ (even $r$ and an arbitrary modulus of continuity $\omega(t)$) on information $T_r(f, X)$ in uniform and in integral metrics are studied, and the extremal problem (1.5.29), (1.5.30) in restricted statement is solved.

*Theorem 2.5.2* ([25]). Let $n \in \mathbb{N}$, $r(r \in \mathbb{N})$ be an even number, $v \in \{0, 1, 2, 3\}$, $X \in (X)_v$ and $\omega(t)$ be an arbitrary modulus of continuity.

Out of all methods of the form (1.5.25) for recovery of functions of the class $W^r H^{\omega}$ on information $T_r(f, X)$, $X \in (X)_v$ the method $\Phi^r(f, X_v; x)$ designed according to the formula (1.5.25) by means of the mesh $X_v$, is optimal both in uniform and in integral metrics.

Furthermore,

$$\Delta_n^v(W^r H^{\omega})_C = \inf_{X \in (X)_v} \Delta(W^r H^{\omega}, X)_C = \sup_{f \in W^r H^{\omega}} \| f(\cdot) - \Phi^r(f, X_v; \cdot) \|_C$$

$$= \frac{D^r(A_v, X_v)_{\infty}}{(r-1)!} \int_0^1 (1 - z)^{r-1} \omega(z \cdot D(A_v, X_v)_{\infty})\mathrm{d}z \tag{1.7.28}$$

and

$$\Delta_n^v(W^r H^{\omega})_{L_p} = \inf_{X \in (X)_v} \Delta(W^r H^{\omega}, X)_{L_p} = \sup_{f \in W^r H^{\omega}} \| f(\cdot) - \Phi^r(f, X_v; \cdot) \|_{L_p}$$

† In [39] the result is given for an arbitrary mesh $\Delta$, while here it is given for a uniform mesh only.

$$= \frac{D^r(A_v, X_v)_\infty}{(r-1)!} \left\{ \int\limits_0^1 \left[ \int\limits_0^1 u^r (1-z)^{r-1} \omega(uz \cdot D(A_v, X_v)_\infty) dz \right]^p dz \right\}^{1/p},$$

$$p > 0. \tag{1.7.29}$$

The supremums in (1.7.28) and (1.7.29) are attained by the function $f_r(\omega, X_v; x)$, where

$$f_r(\omega, X; x) = \frac{1}{(r-1)!} \int\limits_0^x (x-t)^{r-1} f_0(\omega, X; t) dt, \qquad x \in [0,1], \tag{1.7.30}$$

$f_0(\omega, X; t)$ being the standard function for $H^\omega$ and for the mesh $X$.

$(A_v, X_v)$ is a mesh pair determined in $(1.6.5_v)$, and $D(A, X)_\infty$ is the uniform deviation of $(A, X)$.

Chapter 4 is devoted to the problem of recovery of functions of two variables of the classes $H^{\omega_1, \omega_2}(D_2)$ and $H^\omega(D_2)$ in uniform metrics on information given on a lattice of nodes in the unit square.

Let us introduce some notation. Let $n$ and $m$ be positive integers, $X = (x_1, x_2, \ldots, x_n)$ and $Y = (y_1, y_2, \ldots, y_m)$ be arbitrary meshes in the segment $[0,1]$. For a lattice $(X, Y)$ of points $M_{ij} = (x_i, y_j)$ $(i = 1, 2, \ldots, n; j = 1, 2, \ldots, m)$ we say that it belongs to the set $(X, Y)_{v,\mu}, v \in \{0, 1, 2, 3\}$, $\mu \in \{0, 1, 2, 3\}$ and write $(X, Y) \in (X, Y)_{v,\mu}$ when $X \in (X)_v$ and $Y \in (Y)_\mu$.

We assume that the information

$$T(f; X, Y) = (f(x_i, y_j))_{i=1, j=1}^{n, m} \tag{1.7.31}$$

is known.

*Theorem 3.1.1* ([64]). let $n$ and $m$ be positive integers, $(X, Y) \in (X, Y)_{0,0}$, and $\omega_1(t)$, $\omega_2(t)$ and $\omega(t)$ be arbitrary moduli of continuity.

Out of all possible methods for recovery of functions of the classes $H^{\omega_1, \omega_2}(D_2)$ and $H^\omega(D_2)$ using the information $T(f; X, Y)$, $(X, Y) \in (X, Y)_{0,0}$ only, the best is

$$\Phi(f; X, Y; x, y) = \sum_{i=1}^{n} \sum_{j=1}^{m} \chi_i(x) \chi_j(y) \cdot f(x_i, y_j), \qquad (x, y) \in D_2 \tag{1.7.32}$$

for the same characteristic functions $\chi_i(x)$ as in (1.7.4), and with the characteristic functions $\chi_j(y)$ of the intervals $[e_{j-1}, e_j)$ $(j = 1, 2, \ldots, m)$, uniquely defined from the mesh $Y$ by

$$e_0 = 0, \qquad e_m = 1, \qquad e_j = \frac{y_j + y_{j+1}}{2} \quad (j = 1, 2, \ldots, m-1). \tag{1.7.33}$$

Further, for each point $(x, y) \in D_2$

$$R(H^{\omega_1 \omega_2}(D_2); X, Y; x, y) = \sup_{f \in H^{\omega_1 \omega_2}(D_2)} |f(x, y) - \Phi(f; X, Y; x, y)|$$

$$= f_{12}(X, Y; x, y) \tag{1.7.34}$$

and

$$R(H^{\omega}(D_2); X, Y; x, y) = \sup_{f \in H^{\omega}(D_2)} |f(x, y) - \Phi(f; X, Y; x, y)| \tag{1.1.11}$$

$$= f_0(X, Y; x, y) \tag{1.7.35}$$

where the supremums in (1.7.34) and (1.7.35) are attained by the standard functions

$$f_{12}(X, Y; x, y) = \sum_{i=1}^{n} \chi_i(x)\omega_1(|x - x_i|) + \sum_{j=1}^{m} \chi_j(y)\omega_2(|y - y_j|) \tag{1.7.36}$$

and

$$f_o(X, Y; x, y) = \sum_{i=1}^{n} \sum_{j=1}^{m} \chi_i(x) \cdot \chi_j(y) \cdot \omega(\rho(M, M_{ij})) \tag{1.7.37}$$

with $\rho(M, M_{ij}) = \sqrt{(x - x_i)^2 + (y - y_j)^2}$, correspondingly.

In the following Theorem 3.1.2 (which we do not reproduce here), the error of the best method for recovery of functions of the classes $H^{\omega_1 \omega_2}(D_2)$ and $H^{\omega}(D_2)$ on information $T(f; X, Y)$, $(X, Y) \in (X, Y)_{0,0}$ in uniform metrics, is calculated.

*Theorem 3.2.1* ([64]). Let $n \in \mathbb{N}$, $m \in \mathbb{N}$, $(v, \mu) \in \{0, 1, 2, 3\} \times \{0, 1, 2, 3\}$, $(X, Y) \in (X, Y)_{v, \mu}$; $\omega_1(t)$, $\omega_2(t)$ and $\omega(t)$ are arbitrary moduli of continuity.

Out of all possible methods for recovery of functions of the classes $H^{\omega_1 \omega_2}(D_2)$ and $H^{\omega}(D_2)$ on information $T(f; X, Y)$, $(X, Y) \in (X, Y)_{v, \mu}$, optimal in the uniform metrics is the method $\Phi(f; X_v, Y_\mu; x, y)$ designed according to the formula (1.7.32) where $X_v$ is determined by (1.6.5$_v$), and $Y_\mu$ by (1.6.5$_\mu$) when $n = m$.

Furthermore,

$$R_{nm}^{v\mu}(H^{\omega_1 \omega_2}(D_2))_{C(D_2)} = \inf_{(X,Y) \in (X,Y)_{v,\mu}} \|R(H^{\omega_1 \omega_2}(D_2); X, Y; .)\|_{C(D_2)}$$

$$= \sup_{f \in H^{\omega_1 \omega_2}(D_2)} \|f(., .) - \Phi(f; X_v, Y_\mu; ., .)\|_{C(D_2)}$$

$$= \|f_{12}(X_v, Y_\mu; ., .)\|_{C(D_2)}$$

$$= \omega_1(D(A_v, X_v)_\infty) + \omega_2(D(A_\mu, Y_\mu)_\infty) \tag{1.7.38}$$

and

$$R_{nm}^{\nu\mu}(H^{\omega}(D_2))_{C(D_2)} = \inf_{(X,Y)\in(X,Y)_{\nu,\mu}} \|R(H^{\omega}(D_2); X, Y; ., .)\|_{C(D_2)}$$

$$= \sup_{f\in H_{\omega}(D_2)} \|f(., .) - \Phi(f; X, Y; ., .)\|_{C(D_2)}$$

$$= \|f_0(X_\nu, Y_\mu; ., .)\|_{C(D_2)}$$

$$= \omega(\sqrt{D^2(A_\nu, X_\nu)_\infty + D^2(A_\mu, Y_\mu)_\infty}). \qquad (1.7.39)$$

At the end of Section 3.2, in Table 3.2.1 the assertions of the theorem are stated explicitely. Each row of this table contains an assertion of Theorem 3.2.1. For comparison we write down only the case $(\nu, \mu) = (2, 2)$ of equalities (1.7.38) and (1.7.40):

$$R_{nm}^{22}(H^{\omega_1\omega_2}(D_2))_{C(D_2)} = \sup_{f\in H^{\omega_1\omega_2}(D_2)} \|f(., .) - \Phi(f; X_2, Y_2; ., .)\|_{C(D_2)}$$

$$= \omega_1(1/2(n-1)) + \omega_2(1/2(m-1)) \qquad (1.7.40)$$

and

$$R_{nm}^{22}(H^{\omega}(D_2))_{C(D_2)} = \sup_{f\in H^{\omega}(D_1)} \|f(., .) - \Phi(f; X_2, Y_2, ., .)\|_{C(D_2)}$$

$$= \omega\left(\frac{1}{2}\sqrt{\frac{1}{(n-1)^2} + \frac{1}{(m-1)^2}}\right). \qquad (1.7.41)$$

Now, we will make a survey of known results concerning the approximation of functions of the classes $H^{\omega_1\omega_2}(D_2)$ and $H^{\omega}(D_2)$ with some classes of splines ([65]–[68], [70], etc.).

*Theorem* (Storchai [68]). Let $\omega_1(t)$ and $\omega_2(t)$ $(\omega_1(t) + \omega_2(t) \neq 0)$ be convex moduli of continuity. Then

$$\sup_{f\in H^{\omega_1\omega_2}(D_2)} \frac{\|f(., .) - \rho_{n,m}(f; ., .)\|_C}{\omega_1(1/2n) + \omega_2(1/2m)} = 1 \qquad (1.7.42)$$

where $\rho_{n,m}(f; x, y)$ is a $\varphi$-spline [68] interpolating the function $f(x, y)$ in the nodes $M_{ij} = (i/n, j/m)$ $(i = 0, 1, 2, \ldots, n; j = 0, 1, 2, \ldots, m)$†.

*Theorem* (Storchai [68]). Let $\omega_1(t)$ and $\omega_2(t)$ $(\omega_1(t) + \omega_2(t) \neq 0)$ be arbitrary moduli of continuity. Then

$$I(\omega_1, \omega_2; n, m) = \sup_{f\in H^{\omega_1\omega_2}(D_2)} \frac{\|f(., .) - \rho_{n,m}(f; ., .)\|_C}{\omega_1(1/2n) + \omega_2(1/2m)} < \frac{3}{2} \qquad (1.7.43)$$

† This lattice has one row and one column more than the lattice $(X_2, Y_2)$.

and

$$\sup_{\omega_1,\omega_2} I(\omega_1,\omega_2;n,m) = \frac{3}{2}.$$

*Theorem* (Storchai [68]). Let $\omega(t) \not\equiv 0$ be a convex modulus of continuity. Then

$$\sup_{f\in H^\omega(D_2)} \frac{\|f(.,.) - \rho_{n,m}(f;.,.)\|_C}{\omega\left(\frac{1}{2}\sqrt{\frac{1}{n^2} + \frac{1}{m^2}}\right)} = 1. \tag{1.7.44}$$

From (1.7.40), (1.7.42) and (1.7.43) it follows that:
(i) (1.7.40) as well as (1.7.45) and (1.7.42) use the same information (only the methods are different);
(ii) estimate (1.7.42) is valid for convex moduli only, but (1.7.40) is valid for arbitrary moduli of continuity also;
(iii) estimate (1.7.40) is better than (1.7.43);
(iv) estimate (1.7.40) has the optimality property.
  Similar things may be said for the estimates (1.7.41) and (1.7.44) too, but in contrast to the previous case, there is no analogue of (1.7.43) for the arbitrary modulus of continuity.
  Chapter 4 is devoted to some (optimal in one sense or another) quadrature formulas for functions of the classes $W^r H^\omega$ ($r = 0, 2, 4, \ldots$; $\omega(t)$ is an arbitrary modulus of continuity) on information $T_r(f;X)$ for the function $f$ made with respect to a free or additionally restricted mesh $X$.
  In Section 4.1 the best quadrature formula for recovery of the integral

$$\int_0^1 f(x)\mathrm{d}x \tag{1.7.45}$$

in the function class $H^\omega$ (under an arbitrary modulus of continuity $\omega(t)$) on information $T_0(f;X)$ is found.
  *Theorem 4.1.1.* Let $n\in\mathbb{N}$, $X\in(X)_0$, and $\omega(t)$ be an arbitrary modulus of continuity. Then, out of all quadrature formulas for recovery of integral (1.7.45) in the class $H^\omega$ on information $T_0(f;X)$, $X\in(X)_0$, the best is the formula

$$\int_0^1 f(x)\mathrm{d}x \approx \sum_{i=1}^n (b_i - b_{i-1})f(x_i) \equiv Q^0(f;X) \tag{1.7.46}$$

with the mesh $B = (b_0, b_1, \ldots, b_n)$ determined in (1.7.5). Furthermore,

$$R(H^\omega, X) = \sup_{f \in H^\omega} \left| \int_0^1 f(x)dx - Q^0(f, X) \right| = L_n(\omega; B, X) \qquad (1.7.47)$$

where $L_n(\omega; B, X)$ is determined in (1.7.2), and the supremum in (1.7.47) is attainable for the standard function $f_0(\omega, X; x)$.

In Section 4.2 an optimal quadrature formula for the class $H^\omega$ on information $T_0(f; X)$, $X \in (X)_v$, $v \in \{0, 1, 2, 3\}$ is found.

*Theorem 4.2.1.* Let $n \in \mathbb{N}$, $v \in \{0, 1, 2, 3\}$, $X \in (X)_v$ and $\omega(t)$ be an arbitrary modulus of continuity. Then, out of all quadrature formulas for recovery of integral (1.7.45) in the class $H^\omega$ on information $T_0(f; X)$, $X \in (X)_v$ the optimal one is the formula

$$\int_0^1 f(x)dx \approx Q^0(f; X_v) \qquad (1.7.48)$$

where $Q^0(f, X_v)$ is determined by (1.7.46), and the mesh $X_v$ by formula $(1.6.5_v)$. Further,

$$R_n^v(H^\omega) = \inf_{X \in (X)_v} R(H^\omega, X) = \sup_{f \in H} \left| \int_0^1 f(x)dx - Q^0(f; X_v) \right|$$

$$= \int_0^1 f_0(\omega, X_v; x)dx - Q^0(f_0; X_v) = \int_0^1 \omega(u \cdot D(A_v, X_v)_\infty)du \qquad (1.7.49)$$

where $f_0(\omega, X_v; x)$ is the standard function, $D(A, X)_\infty$ is the uniform deviation of the mesh pair $(A, X)$ and $(A_v, X_v)$ $(v \in \{0, 1, 2, 3\})$, determined by $(1.6.5_v)$. The case $v = 0$ of Theorem 4.2.1 is the following known result of Korneichuk:

*Corollary 4.2.1* (Korneichuk [19]). Let $n \in \mathbb{N}$, and $\omega(t)$ be an arbitrary modulus of continuity. Then, out of all formulas for recovery of integral (1.7.45) in the class $H^\omega$ on information $T_0(f; X)$, $X \in (X)_0$, the optimal one is the quadrature formula of the middle rectangles

$$\int_0^1 f(x)dx \approx \frac{1}{n} \sum_{i=1}^{n} f\left(\frac{2i-1}{2n}\right) \equiv Q^0(f; X_0) \qquad (1.7.50)$$

with the error

$$R_n^0(H^\omega) = \int_0^1 \omega\left(\frac{u}{2n}\right)du. \tag{1.7.51}$$

Taking $\omega(t) = t^\alpha, 0 < \alpha \leqq 1$, Corollary 4.2.1 implies the yet-to-be mentioned result of Turezkii [18].

In the case $v = 1$, Theorem 4.2.1 yields:

*Corollary 4.2.3.* Let $n \in \mathbb{N}$ and $\omega(t)$ be an arbitrary modulus of continuity. Then, out of all quadrature formulas for recovery of integral (1.7.45) in the class $H^\omega$ on information $T_0(f; X)$, $X \in (X)_1$, the optimal one is

$$\int_0^1 f(x)dx \approx \frac{2}{2n-1}\left[\frac{f(0)}{2} + \sum_{i=2}^{n} f\left(\frac{2i-2}{2n-1}\right)\right] \equiv Q^0(f; X_1) \tag{1.7.52}$$

with the error

$$R_n^1(H^\omega) = \int_0^1 \omega(u/(2n-1))du. \tag{1.7.53}$$

From (1.7.53) it follows that for $\omega(t) = t^\alpha$, $0 < \alpha \leqq 1$

$$R_n^1(H^\alpha) = \frac{1}{(\alpha+1)(2n-1)^\alpha} \tag{1.7.54}$$

and from (1.7.54) for $\alpha = 1$ it follows that

$$R_n^1(W_\infty^1) = 1/2(2n-1) \tag{1.7.55}$$

since $H^1 = W_\infty^1$.

Following Nikol'skii ([1], p 26), let us denote

$$W_0^1[1; 0, 1] = \{f(x): f \in W_\infty^1, f(0) = 0\}.$$

From Corollary 4.2.3 for $n = m + 1$ the following assertion can be obtained:

*Corollary 4.2.4* (Nikol'skii [1]), p 90). Let $n = m + 1$, then, out of all quadrature formulas for recovery of integral (1.7.45) in the class $W_0^1[1; 0, 1]$

on information $T_0(f;X)$, $X \in (X)_1$ the optimal one is

$$\int_0^1 f(x)dx \approx \frac{2}{2m+1} \sum_{i=2}^{m+1} f\left(\frac{2i-2}{2m+1}\right) \tag{1.7.56}$$

with the error

$$R_{m+1}^1(W_0^1[1;0,1]) = \frac{1}{2(2m+1)}. \tag{1.7.57}$$

For $v = 2$ Theorem 4.2.1 yields.

*Corollary 4.2.5.* Let $n = 2, 3, 4, \ldots$, and let $\omega(t)$ be an arbitrary modulus of continuity. Then, out of all quadrature formulas for recovery of integral (1.7.45) in the class $H^\omega$ on information $T_0(f;X)$, $X \in (X)_2$ the optimal one is the trapezoid formula

$$\int_0^1 f(x)dx \approx \frac{1}{n-1}\left[\frac{f(0)+f(1)}{2} + \sum_{i=2}^{n-1} f\left(\frac{i-1}{n-1}\right)\right] \equiv Q^0(f;X_2) \tag{1.7.58}$$

with the error

$$R_n^2(H^\omega) = \int_0^1 \omega\left(\frac{u}{2n-2}\right)du. \tag{1.7.59}$$

A direct implication of this assertion for $\omega(t) = t^\alpha$, $0 < \alpha \leqq 1$ is Turezkii's result [18]; further, for $\alpha = 1$ we have the classical estimation

$$R_n^2(H^1) = R_n^2(W_\infty^1) = 1/4(n-1).$$

Finally, Theorem 4.2.1 for $v = 3$ yields:

*Corollary 4.2.6.* Let $n \in \mathbb{N}$, and $\omega(t)$ be an arbitrary modulus of continuity. Then, out of all quadrature formulas for recovery of integral (1.7.45) in the class $H^\omega$ on information $T_0(f;X)$, $X \in (X)_3$ the optimal one is

$$\int_0^1 f(x)dx \approx \frac{2}{2n-1}\left[\frac{f(1)}{2} + \sum_{i=1}^{n-1} f\left(\frac{2i-1}{2n-1}\right)\right] \equiv Q^0(f,X_3) \tag{1.7.60}$$

with the error

$$R_n^3(H^\omega) = \int_0^1 \omega(u/(2n-1))\,du. \tag{1.7.61}$$

Section 4.3 is an auxiliary one. It aims to stress the influence of the restriction $x_1 = 0$ or $x_n = 1$, or (finally) $x_1 = 0$, $x_n = 1$ and $n > 1$ on the errors of the optimal quadrature formulas for individual formulas, considered in the previous section, i.e. the influence of the endpoints. That is to say, when the endpoints $x_1 = 0$ and $x_n = 1$ $(n > 1)$ of the mesh $X$ are fixed, then the error of the optimal quadrature formula increases for each individual function $f \in R[0, 1]$.

Niederreiter [31] had shown that for each mesh $X \in (X)_0$ and for each function $f \in C$, the estimate

$$\left| \int_0^1 f(x)\,dx - \frac{1}{n} \sum_{i=1}^n f(x_i) \right| \leqq \omega(f; D(A_0, X)_\infty)$$

with $A_0 = (0, 1/n, 2/n, \ldots, n/n)$ holds, where $D(A, X)_\infty$ is the uniform deviation of the mesh pair $(A, X)$.

Proinov [20] proposed a generalization of Niederreiter's result for arbitrary non-negative weights, showing that if $f \in C$ and $X \in (X)_0$, then

$$\left| \int_0^1 f(x)\,dx - \sum_{i=1}^n p_i f(x_i) \right| \leqq \omega(f; D(A, X)_\infty)$$

with $p_i \geqq 0$, $\sum_{i=1}^n p_k = 1$, $a_0 = 0$, $a_i = \sum_{k=1}^i p_k$ $(i = 1, \ldots, n)$, $A = (a_0, a_1, \ldots, a_n)$.

Further generalizations of the results of Niederreiter for some classes of functions with $r$th$(r \in \mathbb{N})$ derivatives are given in the papers of Proinov and Kirov [32], [33], Christov [34] and Totkov and Baselkov [35].

In Section 4.4 the optimization problem for the quadrature formula of Niederreiter type for the function classes $W^r H^\omega$ is considered. Proofs of results announced in [26] are presented (in an elaborate and complemented form).

Let $n \in \mathbb{N}$, $A = (a_0, a_1, \ldots, a_n)$, $X = (x_1, x_2, \ldots, x_n)$, $(A, X) \in (A, X)_0$, $\omega(t)$ be an arbitrary modulus of continuity, $r \in \mathbb{N}$ and $f \in W^r H^\omega$.

A quadrature formula

$$\int_0^1 f(x)\,dx \approx Q^r(f; A, X) \tag{1.7.62}$$

with

$$Q^r(f; A, X) = \sum_{i=1}^{n} \sum_{k=1}^{n} \lambda_{ik} \cdot f^{(k)}(x_i) \tag{1.7.63}$$

and

$$\lambda_{ik} = \frac{(a_i - x_i)^{k+1} - (a_{i-1} - x_i)^{k+1}}{(k+1)!} \quad (i = 1, \ldots, n; k = 0, ., r) \tag{1.7.64}$$

is said to be a quadrature formula of Niederreiter type.

We denote

$$R_r(f; A, X) = \int_0^1 f(x)dx - Q^r(f; A, X). \tag{1.7.65}$$

Let $B = (b_0, b_1, \ldots, b_n)$ be the mesh, uniquely determined by the mesh $X$ by means of formulas (1.7.5). Then, we consider the quadrature formula

$$\int_0^1 f(x)dx \approx Q^r(f; B, X) \tag{1.7.66}$$

with the error term

$$R_r(f; B, X) = \int_0^1 f(x)dx - Q^r(f; B, X) \equiv R_r(f; X). \tag{1.7.67}$$

*Lemma 4.4.1.* Let $n \in \mathbb{N}$, $r(r \in \mathbb{N})$ be an even number, $X \in (X)_0$ and $\omega(t)$ be an arbitrary modulus of continuity. Then

$$\Delta(W^r H^\omega; X) \overset{def}{=} \sup_{f \in W^r H^\omega} |R_r(f; B, X)| = R_r(f_r; B, X) = L_n(\varphi; B, X) \tag{1.7.68}$$

where $L_n(\varphi; A, X)$ is determined in (1.7.2), while

$$\varphi(u) = \frac{u^r}{(r-1)!} \int_0^1 (1-z)^{r-1} \omega(zu)dz, \qquad u \in [0, 1] \tag{1.7.69}$$

and the function $f_r(x) = f_r(\omega, X; x)$ is defined in (1.7.30).

Further, if for a mesh pair $(A, x) \in (A, X)_0$ the equality

$$\sup_{f \in W^r H^\omega} |R_r(f; A, X)| = L_n(\varphi; A, X) \tag{1.7.70}$$

holds, then the equality

$$\sup_{f \in W^r H^\omega} |R_r(f; A, X)| \geq \sup_{f \in W^r H^\omega} |R_r(f; B, X)| \tag{1.7.71}$$

or, which is the same,

$$\sup_{f \in W^r H^\omega} |R_r(f; B, X)| = L_n(\varphi; B, X) = \inf_A L_n(\varphi; A, X) \tag{1.7.72}$$

holds. This means that quadrature formula (1.7.66) is the best among all the quadrature formulas (1.7.62) of Niderreiter type.

*Theorem 4.4.1.* Let $n \in \mathbb{N}$, $r(r \in \mathbb{N})$ be an even number, $v \in \{0, 1, 2, 3\}$, and $\omega(t)$ be an arbitrary modulus of continuity. Then, out of all quadrature formulas of Niderreiter type using the information $T_r(f, X), X \in (X)_v$ only, the optimal in the class $W^r H^\omega$ is the formula

$$\int_0^1 f(x)\mathrm{d}x = Q^r(f; A_v, X_v) + R_r(f; A_v, X_v) \tag{1.7.73}$$

constructed according to formulas (1.7.62–(1.7.65) by means of the mesh pair $(A_v. X_v)$, defined in $(1.6.5_v)$. Further,

$$\Delta_n^v(W^r H^\omega) \overset{def}{=} \inf_{X \in (X)_v} \Delta(W^r H^\omega; X)$$

$$= \sup_{f \in W^r H^\omega} \left| \int_0^1 f(x)\mathrm{d}x - Q^r(f; A_v, X_v) \right|$$

$$= \int_0^1 f_r(\omega, X_v; x)\mathrm{d}x - Q^r(f_r; A_v, X_v)$$

$$= \frac{D^r(A_v, X_v)_\infty}{(r-1)!} \cdot \int_0^1 \int_0^1 u^r(1-z)^{r-1} \cdot \omega(z \cdot u \cdot D(A_v, X_v)_\infty)\mathrm{d}z\, \mathrm{d}u \tag{1.7.74}$$

where $f_r(x) = f_r(\omega, X; x)$ is defined in (1.7.30), $(A_v, X_v)$ in $(1.6.5_v)$, and $D(A, X)_\infty$ is the uniform deviation of the mesh pair $(A, X)$.

From (1.7.74) it follows that the function $f_r(\omega, X_v, x)$ is extremal.

Further, in the form of corollaries, the assertions of the theorem for $v = 0, 1, 2, 3$ are given. Here we reproduce only the case $v = 0$.

*Corollary 4.4.2.* Let $n \in \mathbb{N}$, $r = 2, 4, 6, \ldots$ and $\omega(t)$ be an arbitrary modulus of continuity. Then, out of all quadrature formulas of Niderreiter type (1.7.62) using only the information $T_r(f; X)$, $X \in (X)_0$ the optimal in the class $W^r H^\omega$ is the formula

$$\int_0^1 f(x)\mathrm{d}x = Q^r(f; A_0, X_0) + R_r(f; A_0, X_0) \tag{1.7.75}$$

where

$$Q^r(f; A_0, X_0) = \frac{1}{n} \sum_{i=1}^{n} \sum_{k=0}^{[r/2]} \frac{f^{(2k)}\left(\dfrac{2i-1}{2n}\right)}{(2n)^k \cdot (2k+1)!}$$

with the error

$$\Delta_n^0(W^r H^\omega) = R_r(f_r; A_0, X_0)$$

$$= \frac{1}{(2n)^r(r-1)!} \int_0^1 \int_0^1 u^r(1-z)^{r-1} \omega\left(\frac{uz}{2n}\right) \mathrm{d}z\, \mathrm{d}u \tag{1.7.76}$$

and with the extremal function $f_r(\omega, X_0; x)$, defined in (1.7.30).

For a quadrature formula

$$\int_0^1 f(x)\mathrm{d}x = Q(f) + R(f)$$

it is said [22] that it has an algebraic degree precision of $m$, if for each algebraical polynomial $P(x)$ of degree $\leq m$, the relation $R(P) = 0$ holds, and if there exists a polynomial $P_1(x)$ of degree $m + 1$ for which $R(P_1) \neq 0$.

*Proposition 4.4.1.* If $n \in \mathbb{N}$, $r = 2, 4, \ldots$, then the algebraic degree of precision of the quadrature formula (1.7.75) is $r + 1$.

In Chapter 5 optimal quadrature formulas for the function classes $H^{\omega_1 \omega_2}(D_2)$ and $H^\omega(D_2)$ are constructed on information $T(f; X, Y)$ under some restrictions on the lattice $(X, Y)$ of nodes in $D_2$. The basic results are announced in [30].

In Section 5.1, on the basis of some lemmas and on the results of Chapter 3, the best quadrature formula is found as follows:

*Theorem 5.1.1* ([30]). Let $n \in \mathbb{N}$, $m \in \mathbb{N}$, $(X, Y) \in (X, Y)_{0,0}$, and $\omega_1(t)$, $\omega_2(t)$ be arbitrary moduli of continuity. Then, from all cubature formulas for approximate calculation of the integral

$$\int_0^1\int_0^1 f(x, y)\mathrm{d}x\,\mathrm{d}y \tag{1.7.77}$$

using only the information $T(f; X, Y)$, $(X, Y) \in (X, Y)_{0,0}$, the best in the functional classes $H^{\omega_1\omega_2}(D_2)$ and $H^{\omega}(D_2)$ is

$$\int_0^1\int_0^1 f(x, y)\mathrm{d}x\,\mathrm{d}y \approx \sum_{i=1}^n \sum_{j=1}^m (b_i - b_{i-1})(e_j - e_{j-1})f(x_i, y_j) \equiv Q(f; X, Y) \tag{1.7.78}$$

$$R(f; X, Y) = \int_0^1\int_0^1 f(x, y)\mathrm{d}x\,\mathrm{d}y - Q(f; X, Y) \tag{1.7.79}$$

with the corresponding error

$$R(H^{\omega_1\omega_2}(D_2); X, Y) = \sup_{f \in H^{\omega_1\omega_2}(D_2)} |R(f; X, Y)|$$

$$= R(f_{12}; X, Y) = \int_0^1\int_0^1 f_{12}(X, Y; x, y)\mathrm{d}x\,\mathrm{d}y \tag{1.7.80}$$

and

$$R(H^{\omega}(D_2); X, Y) = \sup_{f \in H^{\omega}(D_1)} |R(f; X, Y)|$$

$$= R(f_0; X, Y) = \int_0^1\int_0^1 f_0(X, Y; x, y)\mathrm{d}x\,\mathrm{d}y \tag{1.7.81}$$

where $f_{12}(X, Y; x, y)$ and $f_0(X, Y; x, y)$ are the standard functions (1.7.36) and (1.7.37), $B = (b_0, b_1, \ldots, b_n)$ and $E = (e_0, e_1, \ldots, e_m)$ are the meshes, defined by formulas (1.7.5) and (1.7.33) correspondingly.

In Section 5.2 the following basic theorem is proved:

*Theorem 5.2.1.* Let $n \in \mathbb{N}$, $m \in \mathbb{N}$, $(v, \mu) \in \{0, 1, 2, 3\} \times \{0, 1, 2, 3\}$, $(X, Y) \in (X, Y)_{v,\mu}$

and $\omega_1(t)$, $\omega_2(t)$ and $\omega(t)$ be arbitrary moduli of continuity. Then, out of all cubature formulas for approximate calculation of integral (1.7.77), the optimal one in the classes $H^{\omega_1\omega_2}(D_2)$ and $H^\omega(D_2)$ on the information $T(f; X, Y)$, $(X, Y) \in (X, Y)_{v,\mu}$ is

$$\int\limits_0^1\int\limits_0^1 f(x, y)\mathrm{d}x\,\mathrm{d}y \approx Q(f; X_v, Y_\mu) \tag{1.7.82}$$

constructed according to (1.7.78) with $(X, Y) = (X_v, Y_\mu)$. Further,

$$R_{nm}^{v\mu}(H^{\omega_1\omega_2}(D_2)) \stackrel{\mathrm{def}}{=} \inf_{(X,Y)\in(X,Y)_{v,\mu}} R(H^{\omega_1\omega_2}(D_2); X, Y) = R(f_{12}; X_v, Y_\mu)$$

$$= \int\limits_0^1 \omega_1(u \cdot D(A_v, X_v)_\infty)\mathrm{d}u + \int\limits_0^1 \omega_2(u \cdot D(A_\mu, Y_\mu)_\infty)\mathrm{d}u \tag{1.7.83}$$

and

$$R_{nm}^{v\mu}(H^\omega(D_2)) \stackrel{\mathrm{def}}{=} \inf_{(X,Y)\in(X,Y)_{v,\mu}} R(H^\omega(D_2); X, Y) = R(f_0; X_v, Y_\mu)$$

$$= \int\limits_0^1\int\limits_0^1 \omega(\sqrt{u^2 \cdot D^2(A_v, X_v)_\infty + v^2 D^2(A_\mu, Y_\mu)_\infty})\mathrm{d}u\,\mathrm{d}v \tag{1.7.84}$$

where $D(A, X)_\infty$ is the uniform deviation of the mesh pair $(A, X)$; $(A_v, X_v)$ is defined by $(1.6.5_v)$ and $(A_\mu, Y_\mu)$ by $(1.6.5_\mu)$ for $n = m$.

The extremal function in the class $H^{\omega_1\omega_2}(D_2)$ is the standard function $f_{12}(X_v, Y_\mu; x, y)$, defined by (1.7.36), and in the class $H^\omega(D_2)$, the standard function $f_0(X_v, Y_\mu; x, y)$, defined in (1.7.37).

The case $(v, \mu) = (0, 0)$ of Theorem 5.2.1 coincides with the known result of Korneichuk [19].

At the end of this section, in table 5.2.1, all these assertions (their total is 16) of Theorem 5.2.1 are presented.

The quasi-splines used up to now, and defined in (1.5.25), in general have (along with their derivatives) discontinuities at the nodes of the mesh (1.5.21) If they are of degree $r$, they have the maximal defect (see e.g. [12], p 7) $r + 1$. In Chapter 6 the approximative properties of another kind of quasi-splines, the rational quasi-splines, are considered having a defect which is not maximal (and even can be made zero).

In Section 6.1 the definition of the notion of basic quasi-spline is given. It is a generalization of the definitions of Rvachev [37], Konovalov [38] and Storchai and Ligun [39].

Let $A$ be the set of all functions $f : \mathbb{R} \to \mathbb{R}$, which are finite (with the segment $[-1, 1]$ as their carrier), even, increasing on the segment $[-1, 0]$ and determining a curvilinear trapezium on the segment with zero area.

For each function $k(x) \in A$ we define [36] a basic quasi-spline $k_m(x)$ of order $m$ by the formulas

$$k_0(x) = \frac{k(x)}{\int_{-1}^{1} k(t)dt}, \qquad x \in \mathbb{R} \tag{1.7.85}$$

and

$$k_m(x) = \int_{2x-1}^{2x+1} k_{m-1}(t)dt, \qquad m \in \mathbb{N}, \qquad x \in \mathbb{R}. \tag{1.7.86}$$

In Section 6.2 some elementary properties of the basic quasi-splines are studied. In particular, it is proved that $k_m(x) \in A$ for each $m \in \mathbb{N}_0$; a partition of the unit is obtained, etc.

In Section 6.3 the convexity properties of the basic quasi-splines is studied and, as a corollary, the following assertion obtained:

For each $m \in \mathbb{N}$ and for every $x \in [0, 1/2]$ the inequality

$$k_m(x) \geqq 1 - x \tag{1.7.87}$$

holds.

This estimate allows us to prove the following basic theorem (implicitely, it is contained in our paper [41]):

*Theorem 6.3.1.* For each $r \in \mathbb{N}_0$ there exists a function $b : [0, 1] \to \mathbb{R}_+$ with the following properties:

(*a*) $b(x)$ is Riemann integrable on the segment $[0, 1]$;

(*b*) $b(x) + b(1 - x) = 1$, $x \in [0, 1]$;

(*c*) $b(0) = 1$;

(*d*) $x^r \cdot b(x) \geqq (1 - x)^r \cdot b(1 - x) \ \forall \ x \in [0, 1/2] \ (0^\circ \overset{def}{=} 1)$.

The function [41]

$$b(x) = \frac{k_m^r(x)}{k_m^r(x) + k_m^r(1 - x)}, \qquad x \in [0, 1] \tag{1.7.88}$$

with $m \in \mathbb{N}$, $r \in \mathbb{N}_0$ has the properties (*a*)–(*d*).

By $A_r$ we denote the set of all functions $b(x)$ satisfying the hypothesis of Theorem 6.3.1, and each element $b(x)$ of this set will bear the name [42] fundamental rational quasi-spline of degree $r$.

In Section 6.4 some properties of the fundamental rational quasi-splines are proven. These properties are the basis of their approximative properties.

*Lemma 6.4.1.* If $r \in \mathbb{N}_0$, $s \in \mathbb{N}_0$ and $s \leq r$, then $A_r \subset A_s$.

*Lemma 6.4.2.* If $r \in \mathbb{N}_0$ and $b \in A_r$, then for each $x \in [0,1]$ and $r \in \mathbb{N}$ the inequality

$$x^{r+1}b(x) + (1-x)^{r+1}b(1-x) \leq 1/2[x^r \cdot b(x) + (1-x)^r \cdot b(1-x)] \quad (1.7.89)$$

holds. For $r = 0$ the corresponding inequality is

$$x \cdot b(x) + (1-x) \cdot b(1-x) \leq 1/2. \quad (1.7.90)$$

*Corollary 6.4.1.* If $r \in \mathbb{N}_0$, $s \in N_0$, $s \leq r + 1$ and $b \in A_r$, then for each $x \in [0,1]$ the inequality

$$x^s \cdot b(x) + (1-x)^s \cdot b(1-x) \leq 1/2^s \quad (1.7.91)$$

holds.

*Lemma 6.4.3.* If $r \in \mathbb{N}_0$, $b \in A_r$ and the function $\theta : [0,1] \to \mathbb{R}_+$ is increasing, then for each $x \in [0,1]$ the inequality

$$x^r \cdot b(x) \cdot \theta(x) + (1-x)^r \cdot b(1-x) \cdot \theta(1-x) \leq \frac{1}{2^{r+1}}[\theta(x) + \theta(1-x)] \quad (1.7.92)$$

holds.

*Corollary 6.4.2.* Let $r \in \mathbb{N}_0$, $b \in A_r$ and the function $\theta : [0,1] \to \mathbb{R}_+$ be increasing and concave. Then, for each $x \in [0,1]$ the inequality

$$x^r \cdot b(x) \cdot \theta(x) + (1-x)^r \cdot b(1-x) \cdot \theta(1-x) \leq 1/2^r \cdot \theta(1/2) \quad (1.7.93)$$

holds.

In each of the inequalities (1.7.89)–(1.7.93) the equality is attained for $x = 1/2$.

We stress that Lemmas 6.4.2 and 6.4.3 and Corollaries 6.4.1 and 6.4.2 are the basic tool. They are contained (some of them in a weaker form or implicitly) in [41], [43] and [44].

Section 6.5 deals with the approximation of functions of one variable by rational quasi-splines in uniform metrics. The basic results are published in [41], [43] and [44].

Let $n \in \mathbb{N}$, and $\tau_n = (x_0, x_1, \ldots, x_n)$ be an arbitrary mesh in the segment $[0,1]$ with nodes

$$0 = x_0 < x_1 < \ldots < x_n = 1.$$

We denote $h_i = x_i - x_{i-1}$, $i = 1, 2, \ldots, n$, $|\tau_n| = \max\limits_{1 \leq i \leq n} h_i$.

By $\{\tau_n\}$ we denote the set of all such meshes $\tau_n$.

Let $r \in \mathbb{N}_0$, $s \in \mathbb{N}_0$, $v \in \mathbb{N}_0$ and $v \leq r$. If for a function $f : [0, 1] \to \mathbb{R}$ there exists, and information is known that

$$T_r = T_r(f, \tau_n) = (f^{(k)}(x_0), \ldots, f^{(k)}(x_n))_{k=0}^r$$

then for each $b \in A_s$ we introduce ([41], [42]) a rational quasi-spline of $v$th degree by the formula

$$\Phi(x) \equiv \Phi(f; x) \equiv \Phi(b, f_v, \tau_n; x)$$

$$= b(u) \cdot S_v(f, x_{i-1}; x) + b(1 - u) \cdot S_v(f, x_i; x) \quad (1.7.94)$$

with $x \in [x_{i-1}, x_i]$ $(i = 1, 2, \ldots, n)$

$$S_v(f, \alpha; x) = \sum_{k=0}^{v} \frac{f^{(k)}(\alpha)}{k!} \cdot (x - \alpha)^k, \qquad \alpha \in [0, 1] \quad (1.7.95)$$

and $u = (x - x_{i-1})/h_i$ $(i = 1, 2, \ldots, n)$.

By $\mathscr{L}(A_s, T_v, \{\tau_n\})$ we denote the set of all rational quasi-splines of degree $v$, defined by formula (1.7.94). We use the shorter denotation $\mathscr{L}_r = \mathscr{L}(A_r, T_r, \{\tau_n\})$ also.

*Theorem 6.5.1* ([41], [43]). Let $r \in \mathbb{N}$, $s \in \mathbb{N}$, $r \leq s + 1$ and $b \in A_s$. Then

$$\inf_{\tau_n} \sup_{f \in W_\infty^r} \|f(.) - \Phi(b, f_{r-1}, \tau_n, .)\|_C = \frac{1}{(2n)^r \cdot r!} \quad (1.7.96)$$

*Theorem 6.5.2* ([41], [43]). Let $r \in \mathbb{N}$, $s \in \mathbb{N}$, $r \leq s + 1$ and $b \in A_s$. Then

$$\inf_{\tau_n} \sup_{f \in W^r C} \|f(.) - \Phi(b, f_{r-1}, \tau_n; .)\|_C = \frac{1}{(2n)^r \cdot r!}. \quad (1.7.97)$$

It is generally accepted that a modulus of continuity $\omega(t)$ $(0 \leq t \leq T)$ is called 'convex' (or, 'convex from above'), when the function $\omega(t)$ is concave on the segment $[0, T]$.

*Theorem 6.5.3.* Let $r \in \mathbb{N}$, $s \in \mathbb{N}$, $r \leq s$, $r$ even, $b \in A_s$ and $\omega(t)$ be a convex modulus of continuity. Then,

$$\sup_{f \in W^r H^\omega} \|f(.) - \Phi(b, f_r, \tau_n; .)\|_C = \frac{|\tau_n|^r}{2^r \cdot (r-1)!} \cdot \int_0^1 (1 - z)^{r-1} \cdot \omega\left(\frac{|\tau_n|}{2} \cdot z\right) dz. \quad (1.7.98)$$

*Corollary 6.5.1* If $r \in \mathbb{N}$, $s \in \mathbb{N}, r \le s$, $r$ even, $b \in A_s$ and $\omega(t)$ is a convex modulus of continuity, then

$$\inf_{\tau_n} \sup_{f \in W^r H^\omega} \| f(.) - \Phi(b, f_r, \tau_n; .) \|_C = \frac{1}{(2n)^r \cdot (r-1)!} \cdot \int_0^1 (1-z)^{r-1} \cdot \omega\left(\frac{z}{2n}\right) dz.$$

$$(1.7.99)$$

This is one of the strongest assertions in approximation by rational quasi-splines.

*Theorem 6.5.4* ([41]). If $n > 1$, $r \in \mathbb{N}$, $s \in \mathbb{N}$, $r$ odd, $r \le s$, $b \in A_s$, and $\omega(t) \not\equiv 0$ $(0 \le t \le 1)$ is a convex modulus of continuity, then

$$\sup_\omega \inf_{\tau_n} \sup_{f \in W^r H^\omega} \frac{(2n)^r \cdot (r-1)! \| f(.) - \Phi(b, f_r, \tau_n; .) \|_C}{\int_0^1 (1-z)^{r-1} \cdot \omega(z/2n) dz} = 1. \quad (1.7.100)$$

*Corollary 6.5.2* ([41]). If $r \in \mathbb{N}$, $s \in \mathbb{N}$, $r \le s$ and $b \in A_s$ then for each individual function $f \in C^r$ the inequality

$$\| f(.) - \Phi(b, f_r, \tau_n; .) \|_C \le \frac{|\tau_n|^r}{2^{r-1} \cdot (r-1))!} \cdot \int_0^1 (1-z)^{r-1} \cdot \omega\left( f^{(r)}; \frac{|\tau_n|}{2} \cdot z \right) dz$$

$$(1.7.101)$$

holds.

*Theorem 6.5.5* ([41]). If $n > 1$, $r \in \mathbb{N}$, $s \in \mathbb{N}$, $r \le s$, $b \in A_s$ and $\omega(t) \not\equiv 0$ is a convex modulus of continuity, then

$$\sup_\omega \inf_{\tau_n} \sup_{f \in W^r H^\omega} \frac{(2n)^r \cdot r! \| f(.) - \Phi(b, f_r, \tau_n; .) \|_C}{\omega(1/2n(r+1))} = 1. \quad (1.7.102)$$

*Corollary 6.5.3* ([41]). If $n > 1$, $r \in \mathbb{N}$, $s \in \mathbb{N}$, $r \le s$, $b \in A_s$, then for each individual function $f \in C^r$ the inequality (with an uniform mesh $\tau_n$)

$$\| f(.) - \Phi(b, f_r, \tau_n; .) \|_C \le \frac{2}{(2n)^r \cdot r!} \cdot \omega\left( f^{(r)}; \frac{1}{2n(r+1)} \right) \quad (1.7.103)$$

holds.

*Theorem 6.5.6.* If $r \in \mathbb{N}$, $s \in \mathbb{N}$, $r \le s$ and $b \in A_s$ then for each individual function $f \in C^r$ the inequality

$$\|f(.) - \Phi(b, f_r, \tau_n; .)\|_C \leq \frac{|\tau_n|^r}{2^{r+1} \cdot (r-1)!}$$

$$\times \max_{0 \leq u \leq 1/2} \int_0^1 (1-z)^{r-1}[\omega(f^{(r)}; |\tau_n| \cdot u \cdot z) + \omega(f^{(r)}; |\tau_n| \cdot z \cdot (1-u))]\mathrm{d}z \quad (1.7.104)$$

holds.

*Corollary 6.5.5.* ([41]). If $r \in \mathbb{N}$, $s \in \mathbb{N}$, $r \leq s$, $b \in A_s$ and $\omega(t)$ is an arbitrary modulus of continuity, then

$$\sup_{f \in W^r H^\omega} \|f(.) - \Phi(b, f_r, \tau_n; .)\|_C$$

$$\leq \frac{|\tau_n|^r}{2^{r+1}(r-1)!} \cdot \int_0^1 (1-x)^{r-1}\left[\omega\left(\frac{|\tau_n|}{2} \cdot z\right) + \omega(|\tau_n| \cdot z)\right]\mathrm{d}z$$

$$\leq \frac{|\tau_n|^r}{2^r \cdot (r-1)!} \cdot \int_0^1 (1-z)^{r-1} \cdot \min\left[\frac{3}{2}\omega\left(\frac{|\tau_n|}{2} \cdot z\right), \omega(|\tau_n| \cdot z)\right]\mathrm{d}z. \quad (1.7.105)$$

Up to this point (in Section 6.5) we have excluded the case $r = 0$. Now we turn to it.

*Corollary 6.5.6.* Let $s \in \mathbb{N}_0$ and $b \in A_s$. Then, if $\omega(t)$ $(0 \leq t \leq 1)$ is an arbitrary modulus of continuity

$$\sup_{f \in H^\omega} \|f(.) - \Phi(b, f_0, \tau_n; .)\|_C \leq \min\left\{\frac{3}{2}\omega\left(\frac{|\tau_n|}{2}\right), \omega(|\tau_n|)\right\} \quad (1.7.106)$$

holds, but if $\omega(t)$ is a convex modulus of continuity, then

$$\sup_{f \in H^\omega} \|f(.) - \Phi(b, f_0, \tau_n; .)\|_C = \omega\left(\frac{|\tau_n|}{2}\right). \quad (1.7.107)$$

Obviously, (1.7.106) and (1.7.107) are analogues of results of Loginov [61] and Malozemov [69] for approximation of functions of the classes $H^\omega$ by rational quasi-splines.

In Section 6.6 approximate properties of the rational quasi-splines in integral metrics, ammounced in [42], are considered.

*Theorem 6.6.1* ([42]). If $f \in C[0.1]$, $1 \leq p < \infty$, $s \in \mathbb{N}_0$ and $\Phi \in \mathscr{L}(A_s, T_0, \{\tau_n\})$, then

$$\|f - \Phi f\|_{L_p[0,1]} \leq \left( \sum_{i=1}^{n} h_i \int_0^1 \omega^p(f; h_i u) du \right)^{1/p}. \qquad (1.7.108)$$

Here, a frequent use of the next proposition is made.

*Lemma 6.5.1.* Let $P$ be an arbitrary non-empty set, $\tau > 0$, $g: P \to [0, \tau]$, there exists $x_0 \in P$ such that $g(x_0) = \inf_{x \in P} g(x)$ and the function $f:[0, \tau) \to \mathbb{R}$ is increasing. Then

$$\inf_{x \in P} f(g(x)) = f(\inf_{x \in P} g(x)) \qquad (1.7.109)$$

*Corollary 6.5.1.* If $1 \leq p < \infty$, $s \in \mathbb{N}_0$, $\Phi \in Z(A_s, T_0, \{\tau_n\})$ and $\omega(t)$ $(0 \leq t \leq 1)$ is an arbitrary modulus of continuity, then

$$\sup_{f \in H^\omega} \|f - \Phi f\|_{L_p[0,1]} \leq \left( \sum_{i=1}^{n} h_i \int_0^1 \omega^p(h_i u) du \right)^{1/p} \qquad (1.7.110)$$

and

$$\inf_{\tau_n} \sup_{f \in H^\omega} \|f - \Phi f\|_{L_p[0,1]} \leq \left( \int_0^1 \omega^p(u/n) du \right)^{1/p}. \qquad (1.7.111)$$

*Corollary 6.6.2.* Let $1 \leq p < \infty$, and $\omega(t)$ $(0 \leq t \leq 1)$ be an arbitrary modulus of continuity. Then the method $\Phi \in Z(A_s, T_0, \{\tau_n\})$ of the rational quasi-splines is optimal (with respect to the order) for recovery of functions of the class $H^\omega$ on the information $T_i(f, \tau_n)$ in $L_p$-metrics, i.e. the inequality

$$R_{n+1}^2(H^\omega)_{L_p} \leq \inf_{\tau_n} \sup_{f \in H^\omega} \|f - \Phi f\|_{L_p} \leq 2 \cdot R_{n+1}^2(H^\omega)_{L_p} \qquad (1.7.112)$$

is valid, where $R_{n+1}^2(H^\omega)_{L_p}$ is the error (2.4.142) of the optimal method $\Phi^0(f, X_2; x)$.

*Theorem 6.6.2* ([42]). If $r \in \mathbb{N}$, $s \in \mathbb{N}$, $r \leq s$, $\Phi \in \mathscr{L}(A_s, T_r, \{\tau_n\})$, $f \in C^r[0, 1]$, $1 \leq p < \infty$, then

$$\|f - \Phi f\|_{L_p[0,1]} \leq \frac{|\tau_n|^r}{2^r \cdot (r-1)!}$$

$$\times \left\{ \sum_{i=1}^{n} h_i \left[ \int_0^1 (1-z)^{r-1} \cdot \|\omega(f^{(r)}; h_i z_0)\|_{L_p[0,1]} dz \right]^p \right\}^{1/p}. \qquad (1.7.113)$$

*Corollary 6.6.3.* If $r \in \mathbb{N}$, $s \in \mathbb{N}, r \leq s$, $\Phi \in \mathscr{L}\ (A_s, T_r, \{\tau_n\})$, $f \in C^r[0,1]$ and $1 \leq p < \infty$, then

$$\|f - \Phi f\|_{L_p[0,1]} \leq \frac{|\tau_n|^r}{2^r \cdot (r-1)!} \cdot \int_0^1 (1-z)^{r-1} \|\omega(f^{(r)}; |\tau_n|z \cdot)\|_{L_p} dz. \quad (1.7.114)$$

*Corollary 6.6.4* ([42]). If $r \in \mathbb{N}_0$, $\Phi \in \mathscr{L}_r$, $f \in C^r$ and $1 \leq p < \infty$, then

$$\|f - \Phi f\|_{Lp[0,1]} \leq \frac{|\tau_n|^r}{2^r \cdot r!} \cdot \|\omega(f^{(r)}; |\tau_n| \cdot)\|_{L_p[0,1]}. \quad (1.7.115)$$

*Corollary 6.6.5.* If $r \in \mathbb{N}$, $s \in \mathbb{N}$, $r \leq s$, $1 \leq p < \infty$, $\Phi \in \mathscr{L}\ (A_s, T_r, \{\tau_n\})$ and $\omega(t)$ is an arbitrary modulus of continuity, then

$$\sup_{f \in W^r H^\omega} \|f - \Phi f\|_{L_p[0,1]} \leq \frac{|\tau_n|^r}{2^r \cdot (r-1)!} \int_0^1 (1-z)^{r-1} \cdot \|\omega(|\tau_n|z \cdot)\|_{L_p[0,1]} dz.$$

$$(1.7.116)$$

*Theorem 6.6.3.* If $s \in \mathbb{N}_0$, $1 \leq p < \infty$, $f : [0,1] \to \mathbb{R}$ is a bounded measurable function (i.e. $f \in M[0,1]$) and $\Phi \in Z\ (A_s, T_0; \{\tau_n\})$, then

$$\|f - \Phi f\|_{L_p} \leq \frac{1}{2}\left[\tau\left(f, 1; \frac{|\tau_n|}{2}\right)_{\infty,p} + \tau(f, 1; |\tau_n|)_{\infty,p}\right] \quad (1.7.117)$$

where $\tau(f, 1; \delta)_{\infty,p}$ is the averaged modulus, introduced by Ivanov [47].

*Corollary 6.6.6* ([42]). If $s \in \mathbb{N}$, $f \in R[0,1]$, $1 \leq p < \infty$ and $\Phi \in \mathscr{L}\ (A_s, T_0; \{\tau_n\})$, then

$$\|f - \Phi f\|_{L_p[0,1]} \leq \frac{1}{2}[\tau(f; |\tau_n|)_p + \tau(f; 2|\tau_n|)_p]$$

$$\leq \min\left\{\frac{3}{2}\tau(f; |\tau_n|)_p, \tau(f; 2 \cdot |\tau_n|)_p\right\} \quad (1.7.118)$$

where $\tau(f, \delta)_p$ is the averaged modulus of smoothness.

*Theorem 6.6.4* ([42]). If $r \in \mathbb{N}$, $s \in \mathbb{N}$, $r \leq s$, $1 \leq p < \infty$, $f \in L_p^r$ and

$\Phi \in Z \ (A_s, T_r, \{\tau_n\})$, then

$$\|f - \Phi f\|_{L_p[0,1]} \leq \frac{|\tau_n|^r}{2^{r+1} \cdot r!} \cdot [\tau(f^{(r)}; |\tau_n|)_p + \tau(f^{(r)}; 2|\tau_n|)_p]. \quad (1.7.119)$$

*Corollary 6.6.7.* If $r \in \mathbb{N}_0$, $s \in \mathbb{N}_0$, $r \leq s$, $1 \leq p < \infty$, $f \in L_p^r[0,1]$ and $\Phi \in \mathscr{L}$ $(A_s, T_r; \{\tau_n\})$, then

$$\| f - \Phi f \|_{L_p[0,1]} \leq \frac{|\tau_n|^r}{2^r \cdot r!} \cdot \frac{3}{2} \tau(f^{(r)}; |\tau_n|)_p. \quad (1.7.120)$$

In Chapter 7 the theory of the atomar function [50], [51], [62], [63] is developed, but here it is built on the basis of the basic quasi-splines, since (in the author's opinion) in this way its approximate properties are made clear along with its connection to the approximation of functions by rational quasi-splines. This chapter is intended to give an answer to an important question for the rational quasi-splines: Does there exist an infinitely many times differentiable fundamental rational quasi-spline?

In Section 7.1 the representation of the derivatives of the basic quasi-splines by the values of a basic quasi-spline of lower order and the expansions of the polynomials on the shifts of such splines is obtained. We give here only two such propositions.

*Theorem 7.1.9* ([52]). If $m = 2, 3, \ldots$ ; $s = 1, 2, \ldots, m-1$ and $x \in [-1, 1]$, then

$$k_m^{(s)}(x) = (-1)^{\sigma([2^{s-1}(x+1)])} \cdot 2^{s(s+1)/2} \cdot k_{m-s}(-1 + 2 \cdot \{2^{s-1}x\}) \quad (1.7.121)$$

where the notation $[y]$ and $\{y\}$ are used for the integer and fractional parts of the number $y$, respectively; $\sigma(p)$ is the sum of the dyadic figures of $p$, and $k_v(x)$ is a basic quasi-spline of order $v$.

*Corollary 7.1.3.* If $m = 2, 3, \ldots$ and $s = 1, 2, \ldots, m-1$ then

$$k_m(x) + (-1)^{s-1} \cdot k_m\left(\frac{1}{2^{s-1}} - x\right) = (-1)^{s-1} + \sum_{v=0}^{s-1} \frac{k_m^{(v)}(1/2^{s-1})}{v!} \cdot \left(x - \frac{1}{2^{s-1}}\right)^v$$

$$(1.7.122)$$

for each $x \in [0, 2^{1-s}]$. Further,

$$k_m^{(s-1)}\left(\frac{1}{2^{s-1}}\right) = -2^{(s-1)s/2}, \quad s = 2, 3, \ldots, m-1; \quad m = 3, 4, \ldots \quad (1.7.123)$$

Therefore, the right hand side of (1.7.122) is a polynomial of order $s - 1$.

Since the system

$$
\left|
\begin{aligned}
&k_m(x) + k_m(1 - x) = 1, && x\in[0,1], && m \geq 1;\\
&k_m(x) - k_m(\tfrac{1}{2} - x) = \tfrac{1}{2} - 2x, && x\in[0,\tfrac{1}{2}], && m \geq 3;\\
&k_m(x) + k_m(\tfrac{1}{4} - x) = 1 + k_m(\tfrac{1}{4}) + x - 4x^2, && x\in[0,\tfrac{1}{4}], && m \geq 4;\\
&\text{-----------------------------------}\\
&k_m(x) + (-1)^{s-1} k_m\!\left(\frac{1}{2^{s-1}} - x\right)\\
&\quad = (-1)^{s-1} + \ldots - 2^{(s-1)s/2}\cdot x^{s-1}, && x\in[0, 2^{1-s}] - 1 \quad m \geq s - 1
\end{aligned}
\right.
\tag{1.7.124}
$$

has a unique solution with respect to

$$
1, x, x^2, \ldots, x^{s-1}
\tag{1.7.125}
$$

in the segment $[0, 1/2^{s-1}]$, then each algebraic polynomial $P(x)$ of degree $\leq s - 1$ can be represented in the segment $[0, 1/2^{s-1}]$ as a linear combination of the functions

$$
k_m(x),\ k_m(\tfrac{1}{2} - x),\ \ldots,\ k_m\!\left(\frac{1}{2^{s-1}} - x\right); \quad m \geq s - 1
\tag{1.7.126}
$$

in a unique way.

In Section 7.2 an interesting property (monotonicity) of the basic quasi-splines is shown.

*Lemma 7.2.1.* Let the functions $f(x)$ and $\varphi(x)$ satisfy the conditions:
(i) $f\in A,\ \varphi\in A$;
(ii) $\varphi(x) \leq f(x),\ x\in[0,\tfrac{1}{2}]$;
(iii) the curves $y = f(x)$ and $y = \varphi(x)\ (-1 \leq x \leq 0)$ are (separately) symmetric with respect to the point $(-\tfrac{1}{2}, \tfrac{1}{2})$. Then

$$
J(x) \overset{def}{=} \int_{2x-1}^{0} (f(x) - \varphi(x))\mathrm{d}x \geq 0, \qquad 0 \leq x \leq \tfrac{1}{2}
\tag{1.7.127}
$$

and

$$
\sup\{J(x): 0 \leq x \leq \tfrac{1}{2}\} = J(\tfrac{1}{4}).
\tag{1.7.128}
$$

*Lemma 7.2.3* ([43]). (*Monotonicity of* $(k_m(x))_{m=1}^{\infty}$.) From the validity of the inequality

$$
k_1(x) \leq k_2(x) \qquad (\text{or,}\quad k_1(x) \geq k_2(x)), \quad 0 \leq x \leq \tfrac{1}{2}
\tag{1.7.129}
$$

the validity of the inequality

$$k_m(x) \leqq k_{m+1}(x) \qquad (\text{or,} \quad k_m(x) \geqq k_{m+1}(x)), \quad 0 \leqq x \leqq \tfrac{1}{2} \qquad (1.7.130)$$

for each $m \in \mathbb{N}$ follows.

If for a sequence of basic quasi-splines

$$k_1(x), k_2(x), \ldots, k_m(x), \ldots, \quad (0 \leqq x \leqq \tfrac{1}{2}) \qquad (1.7.131)$$

condition (1.7.129) is fulfilled, we say [43] that it is increasing (or decreasing). A sequence (1.7.131) is said to be monotonic, if it is increasing, or decreasing.

*Corollary 7.2.3.* If a sequence (1.7.131) is monotonic, then it is convergent in the normed linear space $C[-1, 1]$ (and hence, in $C(\mathbb{R})$).

It is a matter for an immediate check that the sequence $(\lambda_m(x))_{m=1}^{\infty}$ of the basic quasi-splines of the function $\lambda_0(x)$: $\lambda_0(x) = 1$ for $|x| < \tfrac{1}{2}$, $\lambda_0(\pm\tfrac{1}{2}) = \tfrac{1}{2}$ and $\lambda_0(x) = 0$ for $|x| > \tfrac{1}{2}$ is decreasing (see (6.1.6)). Therefore, there exists a continuous function $\lambda(x)$, determined by

$$\lambda(x) = \lim_{m \to \infty} \lambda_m(x), \qquad x \in \mathbb{R}. \qquad (1.7.132)$$

The sequence $(\delta_m(x))_{m=1}^{\infty}$ of the basic quasi-splines of the function $\delta_0(x) = \tfrac{1}{2}$ for $|x| < 1$ and $\delta_0(x) = 0$ for $|x| \geq 1$ is increasing (see (6.1.10)), and hence there exists a continuous function

$$\delta(x) = \lim_{m \to \infty} \delta_m(x), \qquad x \in \mathbb{R}. \qquad (1.7.133)$$

On the other hand, the following proposition holds:

*Lemma 7.2.2.* If $m$ and $p$ are positive integers, then for each sequence $(k_m(x))_{m=1}^{\infty}$ of basic quasi-splines

$$\delta_p(x) \leqq k_{m+p}(x) \leqq \lambda_p(x), \qquad 0 \leqq x \leqq \tfrac{1}{2}. \qquad (1.7.134)$$

Therefore,

$$\delta(x) \leqq \lambda(x), \qquad 0 \leqq x \leqq \tfrac{1}{2}. \qquad (1.7.135)$$

It is natural to put the question: Can one assert that each sequence (1.7.131) of basic quasi-splines is convergent in $C(\mathbb{R})$, i.e.

$$f(x) := \lim_{m \to \infty} k_m(x), \qquad x \in \mathbb{R}? \qquad (1.7.136)$$

Let us assume that the limit (1.7.136) exists. Then (according to the elementary properties of the basic quasi-splines) it satisfies the following conditions:

(i)  $f(x) = 0$, $x \in \mathbb{R} \setminus (-1, 1)$;
(ii)  $f(-x) = f(x)$, $x \in \mathbb{R}$;
(iii)  $-1 \leq x_1 < x_2 \leq 0 \Rightarrow f(x_1) \leq f(x_2)$;
(iv)  $f(0) = 1$;

$$(v) \quad f(x) = \int_{2x-1}^{2x+1} f(t)\,dt, \quad x \in \mathbb{R}.$$

Thus we arrive at the following.

*Problem C.* Find all real functions $f : \mathbb{R} \to \mathbb{R}$ which satisfy conditions (i)–(v).

In Section 7.3 we study the properties of the solutions of Problem C.

*Lemma 7.3.1.* If a function $f(x)$ is a solution of Problem C, then it belongs to the class $A$ and it coincides with each of its basic quasi-splines.

This lemma allows immediate transfer of the elementary properties of the basic quasi-splines to the solutions of Problem C also.

The existence of solutions of Problem C follows from (1.7.132) and (1.7.133), and the unicity of the solution follows from (1.7.135) and the following:

*Theorem 7.3.1.* If $\varphi(x)$ and $f(x)$ are two solutions of Problem C, satisfying the condition

$$\varphi(x) \leq f(x), \qquad 0 \leq x \leq \tfrac{1}{2}, \tag{1.7.137}$$

then they coincide: $\varphi(x) = f(x)$, $\forall x \in \mathbb{R}$.

Indeed, from (1.7.132)–(1.7.137) it follows that

$$\delta(x) \leq f(x) \leq \lambda(x), \qquad 0 \leq x \leq \tfrac{1}{2}$$

and by Theorem 7.3.1 $\delta(x) = f(x) = \lambda(x)$ for $x \in \mathbb{R}$.

The solution of Problem C further will be denoted by $\lambda(x)$ too. Now we can answer the question (1.7.136).

*Theorem 7.3.4.* Each sequence of basic quasi-splines $(k_m(x))_{m=1}^{\infty}$ is uniformly convergent in $\mathbb{R}$ to $\lambda(x)$:

$$\lim_{m \to \infty} k_m(x) = \lambda(x), \qquad x \in \mathbb{R}. \tag{1.7.138}$$

*Theorem 7.3.5.* For function $k(x) \in A$ the inequalities

$$\|\lambda(.) - k_{m-1}(.)\|_{C[-1,1]} \leq \lambda_m(\tfrac{1}{4}) - \delta_m(\tfrac{1}{4}), \qquad m \in \mathbb{N}; \tag{1.7.139}$$

$$\|\lambda(.) - \lambda_m(.)\|_{C[-1,1]} \leq \lambda_m(\tfrac{1}{4}) - \delta_m(\tfrac{1}{4}), \qquad m \in \mathbb{N}_0; \tag{1.7.140}$$

$$\|\lambda(.) - \delta_m(.)\|_{C[-1,1]} \leq \lambda_m(\tfrac{1}{4}) - \delta_m(\tfrac{1}{4}), \qquad m \in \mathbb{N}_0 \qquad (1.7.141)$$

are satisfied.

*Theorem 7.3.6.* For each function $k(x) \in A$ the inequalities

$$\|\lambda(.) - k_{m+1}(.)\|_{C[-1,1]} \leq \frac{1}{2^{m+1}}, \qquad m = 2, 3, \ldots; \qquad (1.7.142)$$

$$\|\lambda(.) - \lambda_{m+1}(.)\|_{C[-1,1]} \leq \frac{1}{2^{m+1}}, \qquad m \in \mathbb{N}_0; \qquad (1.7.143)$$

$$\|\lambda(.) - \delta_m(.)\|_{C[-1,1]} \leq \frac{1}{2^{m+1}}, \qquad m = \mathbb{N}_0 \qquad (1.7.144)$$

are satisfied.

Both theorems and the inequalities (1.7.·34) explain why it is important to study in detail the basic quasi-splines of functions (6.1.3) and (6.1.7).

*Theorem 7.3.7.* The function $\lambda(x)$ is strictly increasing in the segment $[-1, 0]$ and strictly decreasing in the segment $[0, 1]$.

*Corollary 7.3.2.* For each $x \in (-1, 1)$

$$\lambda(x) > 0. \qquad (1.7.145)$$

*Corollary 7.3.3.* The function $\lambda(x)$ is strictly convex on the segments $[-1, -\tfrac{1}{2}]$, $[\tfrac{1}{2}, 1]$ and strictly concave in the segment $[-\tfrac{1}{2}, \tfrac{1}{2}]$.

Furthermore, various properties of $\lambda(x)$ are obtained, using the properties of the basic quasi-splines, since $\lambda(x)$ as a solution of Problem C, is a basic quasi-spline (see Lemma 7.3.1). In particular, it is shown that $\lambda(x)$ is non-analytic in any interval $(a, b) \subset (-1, 1)$, $a < b$, although $\lambda(x) \in C^\infty(\mathbb{R})$. All zeroes of each derivative $\lambda^{(s)}(x)$ of $\lambda(x)$ are found located on the segment $[-1, 1]$. Using inequalities (1.7.139)–(1.7.144), an algorithm for tabulation of $\lambda(x)$ is constructed.

The integral representation

$$\lambda(x) = \frac{1}{\pi} \int_0^{+\infty} \cos(tx) \cdot \prod_{k=1}^{\infty} \frac{\sin(t2^{-k})}{t2^{-k}} \, dt, \qquad x \in \mathbb{R} \qquad (1.7.146)$$

shows that $\lambda(x)$ coincides with the atomar function $up(x)$ of Rvachev and Rvachev [50] (see also [51], p 109) obtained as a finite solution of a linear functional-differential equation of the form

$$y^{(n)}(x) + a_1 y^{(n-1)}(x) + \cdots + a_n y(x) = \sum_{k=1}^{M} b_k \cdot y(ax - b_k) \qquad (1.7.147)$$

with $|a| > 1$, $a_i, b_k \in \mathbb{R}$.

We can also mention the relation

$$\lambda(x) = \int_{2x-1}^{2x+1} \lambda(t)\,dt, \qquad x \in \mathbb{R} \tag{1.7.148}$$

obtained there.

Especially useful is Lemma 7.3.1 asserting that the atomar function $\lambda(x)$ coincides with each of its basic quasi-splines. This means that on the basis of the fundamental rational quasi-splines $\lambda(x)$ can be taken (besides, $\lambda(x) \in C^\infty(\mathbb{R})!$):

For each $s \in \mathbb{N}$

$$b(x) = \frac{\lambda^s(x)}{\lambda^s(x) + \lambda^s(1-x)}, \qquad x \in [0,1]. \tag{1.7.149}$$

Under this choice of $b(x) \in A_s$ all the investigations made in Chapter 6 are valid.

# 2

# Recovery of functions
# of one variable

The present chapter deals with some methods of recovery of functions of the classes $W^r H^\omega$ using the information $T_r(f, X)$ in uniform and integral metrics. We are going to study methods optimaly only in the sence that they enable the solution of some of the extremal problems (1.5.15) or (1.5.30).

The basic results have been reported at the International Conference on Constructive Theory of Functions, held in Varna in 1984 and at the Anniversary Scientific Session of the Higher Technical School A Kanchev, held in Russe on October 4–6, 1984. They are published in [24] and [25].

## 2.1   Optimization theorem

We will prove a proposition referred to later as the optimization theorem. It is announced in [28] and it is of importance for further considerations.

### Theorem 2.1.1

*If the function $\varphi:[0, 1] \to \mathbb{R}$ is (strictly) increasing, then for each pair of meshes $(A, X) \in (A, X)_v$ the following inequality*

$$L_n(\varphi; A, X) \geqq L_n(\varphi; A_v, X_v) \tag{2.1.1}$$

*holds, where for each pair of meshes $(A, X) \in (A, X)_0$*

$$L_n(\varphi; A, X) \stackrel{\text{def}}{=} \sum_{i=1}^{n} \int_{a_{i-1}}^{a_i} \varphi(|x - x_i|)\,\mathrm{d}x \tag{2.1.2}$$

*and the pair of meshes $(A_v, X_v)$ are defined in $(1.6.5_v)$ $(v = 0, 1, 2, 3)$.*

*Moreover, the equality (2.1.1) holds (only) for the mesh pair $(A_v, X_v)$ and*

$$L_n(\varphi, A_v, X_v) = \int_0^1 \varphi(u \cdot D(A_v, X_v)_\infty)du \quad (v = 0, 1, 2, 3) \qquad (2.1.3)$$

*or in a more elaborate form*

$$L_n(\varphi; A_0, X_0) = \int_0^1 \varphi(u/2n)du \qquad (2.1.4_0)$$

$$L_n(\varphi; A_1, X_1) = \int_0^1 \varphi(u/(2n-1))du \qquad (2.1.4_1)$$

$$L_n(\varphi; A_2, X_2) = \int_0^1 \varphi(u/(2n-2))du \qquad (2.1.4_2)$$

$$L_n(\varphi; A_3, X_3) = \int_0^1 \varphi(u/(2n-1))du. \qquad (2.1.4_3)$$

### Proof

First we are going to prove the assertion of the theorem under the following additional assumption of the mesh pair $(A, X)$: for each $i = 1, 2, \ldots, n$ the inequalities

$$a_{i-1} \leqq x_i \leqq a_i \qquad (2.1.5)$$

hold.

From (2.1.5) it follows that every mesh pair $(A, X) \in (A, X)_0$ might be (2.1.2) written in the form

$$L_n(\varphi; A, X) = \sum_{i=1}^{n} [\Phi(a_i - x_i) + \Phi(x_i - a_{i-1})] \qquad (2.1.6_0)$$

where

$$\Phi(x) = \int_0^x \varphi(t)dt, \qquad x \in [0, 1]. \qquad (2.1.7)$$

From the condition for the function $\varphi:[0,1]\to\mathbb{R}$ to be (strictly) increasing it follows that function (2.1.7) is (strictly) convex in the segment $[0,1]$.

If $(A,X)\in(A,X)_v$ $(v=1,2,3)$, then $(2.1.6_0)$ is of the form

$$L_n(\varphi;A,X)=\Phi(a_1)+\sum_{i=z}^{n}[\Phi(a_i-x_i)+\Phi(x_i-a_{i-1})],\quad v=1;\quad (2.1.6_1)$$

$$L_n(\varphi;A,X)=\Phi(a_1)+\sum_{i=2}^{n-1}[\Phi(a_i-x_i)+\Phi(x_i-a_{i-1})]+\Phi(1-a_{n-1}),\quad v=2;$$

$$(2.1.6_2)$$

$$L_n(\varphi;A,X)=\sum_{i=1}^{n-1}[\Phi(a_i-x_i)+\Phi(x_i-a_{i-1})]+\Phi(1-a_{n-1}),\quad v=3.\quad (2.1.6_3)$$

From $(6_v)$ $(v=0,1,2,3)$, by using the classical Jensen inequality for a (strictly) convex function (2.1.7) we obtain the following inequality:

For the case $(A,X)\in(A,X)_0$

$$L_n(\varphi;A,X)\geqq 2n\Phi\left(\frac{1}{2n}\right)=\int_0^1\varphi(u/2n)du\qquad (2.1.8_0)$$

since according to (1.6.4)

$$\sum_{i=1}^{n}[(a_i-x_i)+(x_i-a_{i-1})]=a_n=1.\qquad (2.1.9_0)$$

The equality in $(2.1.8_0)$ holds (only) if the following equalities are satisfied:

$$x_1=a_i-x_i=x_i-a_{i-1}\quad (i=1,2,\ldots,n).\qquad (2.1.10_0)$$

For the case $(A,X)\in(A,X)_1$

$$L_n(\varphi;A,X)\geqq(2n-1)\Phi(1/(2n-1))=\int_0^1\varphi(u/(2n-1))du\qquad (2.1.8_1)$$

holds since according to (1.6.4)

$$a_1+\sum_{i=2}^{n}[(a_i-x_i)+(x_i-a_{i-1})]=a_n=1.\qquad (2.1.9_1)$$

The equality in $(2.1.8_1)$ holds (only) if the following equalities

$$a_1 = a_i - x_i = x_i - a_{i-1} \quad (i = 2, 3, \ldots, n) \qquad (2.1.9_1)$$

are satisfied.

In the case $(A, X) \in (A, X)_2$

$$L_n(\varphi; A, X) \geq (2n - 2)\Phi(1/(2n - 2)) = \int_0^1 \varphi(u/(2n - 2))du \qquad (2.1.8_2)$$

holds, since according to (1.6.4) and $X \in (X)_2$

$$a_1 + \sum_{i=2}^{n-1} [(a_i + x_i) + (x_i - a_{i-1})] + x_n - a_{n-1} = x_n = 1. \qquad (2.1.9_2)$$

The equality in $(2.1.8_2)$ holds if the following equalities are satisfied:

$$a_1 = a_i - x_i = x_i - a_{i-1} = 1 - a_{n-1} \quad (i = 2, 3, \ldots, n - 1). \qquad (2.1.10_2)$$

In the case $(A, X) \in (A, X)_3$

$$L_n(\varphi; A, X) \geq (2n - 1)\Phi(1/2n - 1)) = \int_0^1 \varphi(u/(2n - 1))du \qquad (2.1.8_3)$$

holds, since according to (1.6.4) and $X \in (X)_3$

$$\sum_{i=1}^{n-1} [(a_i - x_i) + (x_i - a_{i-1})] + x_n - a_{n-1} = x_n = 1. \qquad (2.1.9_3)$$

The equality in $(2.1.8_3)$ holds (only) if the following equalities

$$a_i - x_i = x_i - a_{i-1} = 1 - a_{n-1} \quad (i = 1, 2, \ldots, n - 1) \qquad (2.1.10_3)$$

are fulfilled.

Since from $(2.1.9_v)$ and $(2.1.10_v)$ $(v = 0, 1, 2, 3)$ these follows $(1.6.5_v)$ $(v = 0, 1, 2, 3)$, then from $(2.1.8_v)$ there follows (2.1.1) and $(2.1.4_v)$ also $v = 0, 1, 2, 3$, i.e. the assertions† of the theorem will be proved if the additional condition is satisfied.

---

† Later we are going to prove it.

Let us now disregard the additional condition. Let it not be satisfied for the mesh pair $(A, X) \in (A, X)_v$, i.e. let at least one of the numbers $i = 1, 2, \ldots, n$ not be fulfilled for at least one of equalities (2.1.5). Then, we construct a new mesh pair $(A, X') \in (A, X)_v$ satisfying the additional condition in the following way: let us put $x'_i = x_i$ for every $i (i = 1, 2, \ldots, n)$ for which the elements of the previous pair of networks satisfies inequalities (2.1.5) and for every $i (i = 1, 2, \ldots, n)$ for which the elements of the previous pair of networks $(A, X)$ does not fulfil at least one of inequalities (2.1.5). We put

$$x'_i = a_{i-1} \qquad \text{if} \quad x_i < a_{i-1}$$

and

$$x'_i = a_i \qquad \text{if} \quad x_i > a_i.$$

Thus a new mesh pair $(A, X') \in (A, X)_v$ is constructed which satisfies the additional condition also. From the proof given above the assertions of the theorem are valid for the new $(A, X')$ constructed. Moreover, for each $x \in [a_{i-1}, a_i]$ $(i = 1, 2, \ldots, n)$ we have: $|x - x_2| \geq |x - x'_i|$.

Hence the increasing behaviour of the function $\varphi : [0, 1] \to \mathbb{R}$ and (2.1.2) implies the inequality

$$L_n(\varphi; A, X) \geqq L_n(\varphi; A, X').$$

Therefore, the assertions of the theorem are true for each mesh pair $(A, X) \in (A, X)_v$ without any reference to the additional condition. (In the above consideration $v$ denotes a fixed element of the set $\{0, 1, 2, 3\}$).

Thus the theorem is completely proved.


**Corollary 2.1.2.** *Under the hypotheses and the denotations of Theorem 2.1.1 the identity*

$$\inf\{L_n(\varphi; A, X) : (A, X) \in (A, X)_v\} = L_n(\varphi; A_v, X_v), \quad v = 0, 1, 2, 3 \quad (2.1.11)$$

*holds.*


Since $(A_v, X_v) \in (A, X)_v$, $v = 0, 1, 2, 3$, then equality (2.1.11) follows from (2.1.1).

Next, let us prove (1.7.3). To this end, first we will show that for each mesh pair $(A, X) \in (A, X)_0$ the equality

$$D(A, X)_p = \left( \sum_{i=1}^{n} \int_{a_{i-1}}^{a_i} |x - x_i|^p \, dx \right)^{1/p}, \qquad p > 0 \qquad (2.1.12)$$

holds.

Indeed, the known identity (see. e.g. [21], p 279)

$$\int_a^b |\xi - x|^p\, dx = \frac{1}{p+1}\left[(b-\xi)\cdot|b-\xi|^p + (\xi-a)\cdot|\xi-a|^p\right], \qquad p \geqq 0 \quad (2.1.13)$$

implies the identity

$$\sum_{i=1}^n \int_{a_{i-1}}^{a_i} |x - x_i|^p\, dx$$

$$= \frac{1}{p+1}\sum_{i=1}^n \left[(a_i - x_i)|a_i - x_i|^p + (x_i - a_{i-1})\cdot|x_i - a_{i-1}|^p\right]. \quad (2.1.14)$$

From (1.6.10) and (1.6.8) the equality

$$D^p(A, X)_p = \int_0^1 |F_n(x) - x|^p\, dx$$

$$= \sum_{i=1}^{n+1} \int_{x_{i-1}}^{x_i} |F_n(x) - x|^p\, dx = \sum_{i=1}^{n+1} \int_{x_{i-1}}^{x_i} |a_{i-1} - x|^p\, dx \quad (2.1.15)$$

follows, where $x_0 = 0$, $x_{n+1} = 1$.

Using (2.1.13) again, from (2.1.15), having in mind that $x_0 = 0$, $x_{n+1} = 1$, $(A, X) \in (A, X)_0$, we obtain the equality

$$D^p(A, X)_p = \frac{1}{p+1}\sum_{i=1}^{n+1} \left[(x_i - a_{i-1})\cdot|x_i - a_{i-1}|^p \right.$$

$$\left. + (a_{i-1} - x_{i-1})\cdot|a_{i-1} - x_{i-1}|^p\right]$$

$$= \frac{1}{p+1}\sum_{i=1}^n \left[(a_i - x_i)\cdot|a_i - x_i|^p + (x_i - a_{i-1})\cdot|x_i - a_{i-1}|^p\right]. \quad (2.1.16)$$

Then, (2.1.15) and (2.1.16) imply (2.1.12).

**Corollary 2.1.3**    *Let $(A, X) \in (A, X)_v$. Then the inequality*

$$D(A, X)_p \geqq D(A_v, X_v)_p, \qquad 0 < p \leqq \infty \qquad (2.1.17)$$

*holds, with equality for the mesh pair $(A_v, X_v)$ defined in $(1.6.5_v)$ only, where $v$ is a fixed element of the set $\{0, 1, 2, 3\}$. If $0 < p < \infty$, then*

$$D(A_0, X_0)_p = 1/2n\sqrt[p]{1+p} \qquad (2.1.17_0)$$

$$D(A_1, X_1)_p = 1/(2n-1)\sqrt[p]{1+p} \qquad (2.1.17_1)$$

$$D(A_2, X_2)_p = 1/(2n-2)\sqrt[p]{1+p} \qquad (2.1.17_2)$$

$$D(A_3, X_3)_p = 1/(2n-1)\sqrt[p]{1+p}. \qquad (2.1.17_3)$$

### Proof

For $0 < p < \infty$ the assertions of Corollary 2.1.3 follow directly from Theorem 2.1.1 and (2.1.12), since the function $\varphi|u| = u^p$, $p > 0$, $u \in [0, 1]$ is strictly increasing.

For $p = \infty$ inequality (2.1.17) follows from the same inequality with $0 < p < \infty$ letting $p \to \infty$, due to (1.6.9)–(1.6.11) and the well known properties of convergent sequences. It remains only to prove that the equality in (2.1.17) (for $p = \infty$) is achieved for the mesh pair $(A_v, X_v)$ only. We will consider the case $v = 1$. (The other cases can be treated in a similar way.)

Let, for a mesh pair $(A, X) \in (A, X)_1$, the equality

$$D(A, X)_\infty = D(A_1, X_1)_\infty = 1/(2n-1) \qquad (2.1.18)$$

hold. Then, (1.6.13), $(A, X) \in (A, X)_1$ and (2.1.18) imply the relations

$$a_0 = x_1 = 0;$$

$$x_i - a_{i-1} \leq |x_i - a_{i-1}| \leq 1/(2n-1) \quad (i = 2, 3, \ldots, n);$$

$$a_i - x_i \leq |a_i - x_i| \leq 1/(2n-1) \quad (i = 1, 2, \ldots, n).$$

Therefore,

$$1 = a_n = a_1 + (x_2 - a_1) + (a_2 - x_2) + \ldots + (a_n - x_n)$$
$$\leq |a_1| + |x_2 - a_1| + |a_2 - x_2| + \ldots + |a_n - x_n|$$
$$\leq (2n-1) \cdot 1/(2n-1) = 1.$$

The last relation implies that in all these inequalities only the $=$ sign holds. In such a way we obtain the equalities

$$a_1 = 1/(2n-1), \qquad x_i - a_{i-1} = 1/(2n-1) \quad (i = 2, 3, \ldots, n)$$

and

$$a_i - x_i = 1/(2n-1) \quad (i = 2, 3, \ldots, n).$$

Hence, the mesh pair $(A, X) \in (A, X)_1$ for which equality (2.1.18) holds is exactly $(A_1, X_1)$.
The proof is complete.

The assertion of Corollary 2.1.3 for $v = 0$ is known (see e.g. [21], pp 62–4).
Now (2.1.3) follows from $(2.1.4_v)$ and $(2.1.17_v)$ letting $p = \infty$.
The assertions of Corollary 2.1.3 can be written in the form

$$\inf_{(A,X) \in (A,X)_v} D(A, X)_p = D(A_v, X_v)_p, \qquad 0 < p \leq \infty, \qquad v = 0, 1, 2, 3,$$

since $(A_v, X_v) \in (A, X)_v$.

## 2.2    The standard function

Let $X = (x_1, x_2, \ldots, x_n)$, $X \in (X)_0$ be an arbitrary mesh of points in the segment $[0, 1]$, $B = (b_0, b_1, \ldots, b_n)$ be the mesh uniquely determined from the mesh $X$ according to formulas (1.5.21), and $\chi_i(x)$ be the characteristic function of the interval $[b_{i-1}, b_i)$ $(i = 1, \ldots, n)$. Let us recall the convention: $[\alpha, \beta) = [\alpha, \beta]$, for $\beta = 1$ and $\alpha \in [0, 1]$ and to note that the characteristic function depends on the mesh $X$ only (the mesh $B$ is uniquely determined by $X$!).

For each mesh $X \in (X)_0$ and for each modulus of continuity $\omega(t)$, $t \in [0, 1]$ we define the standard function $f_0(\omega, X; x) \equiv f_0(x)$ by the formula

$$f_0(x) \equiv f_0(\omega, X; x) = \sum_{i=1}^{n} \chi_i(x) \cdot \omega(|x - x_i|), \qquad x \in [0, 1]. \qquad (2.2.1)$$

Let us introduce the function class

$$H_0^{\omega} = \{ f(x) : f \in H^{\omega}, f(x_i) = 0, i = 1, 2, \ldots, n \}$$
$$= \{ f(x) : f \in H^{\omega}, T_0(f, X) = 0 \}. \qquad (2.2.2)$$

## Lemma 2.2.1

*Let $X \in (X)_0$ be an arbitrary mesh, and $\omega(t)$ ($t \in [0, 1]$) be an arbitrary modulus of continuity. Then the standard function $f_o(x) = f_0(\omega, X; x)$, defined in (2.2.1) has the following properties:*

*(i)* $f_0(x) \geqq 0$ for each $x \in [0, 1]$;

*(ii)* $f_0(x_i) = 0$ for each node $x_i$ $(i = 1, 2, \ldots, n)$ of the mesh $X$;

*(iii)* $f_0(x) \in C[0, 1]$;

*(iv)* $f_0(x) \in H^\omega$;

*(v)* $f_0(x) \in H_0^\omega$;

*(vi)* For each function $f \in H_0^\omega$ and for every $x \in [0, 1]$

$$f(x) \leqq f_o(x), \tag{2.2.3}$$

*that is*

$$\sup\{f(x) : f \in H_0^\omega\} = f_0(x), \qquad x \in [0, 1]. \tag{2.2.4}$$

## Proof

Property (i) follows directly from Definition (2.2.1) since both the characteristic function and the modulus of continuity are non-negative functions.

Since $X \in (X)_0$, then according to (1.5.21)

$$b_0 = 0 \leqq x_1$$

$$b_{i-1} = \frac{x_{i-1} + x_i}{2} < x_i \quad (i = 2, 3, \ldots, n)$$

$$x_i < \frac{x_i + x_{i+1}}{2} = b_i \quad (i = 1, 2, \ldots, n-1)$$

$$x_n \leqq b_n = 1$$

i.e. always $x_i \in [b_{i-1}, b_i)$ $(i = 1, 2, \ldots, n)$. According to (1.5.22) this implies $\chi_i(x_i) = 1$, and $\chi_i(x_k) = 0$ for $k \neq i$. Hence for each node $x_i$ $(i = 1, 2, \ldots, n)$ of the mesh $X$ we have

$$f_0(x_i) = \sum_{k=0}^{n} \chi_k(x_i) \cdot \omega(|x_i - x_k|)$$

$$= 0 \cdot \omega(|x_i - x_1|) + \ldots + 0 \cdot \omega(|x_i - x_{i-1}|) + 1 \cdot \omega(|x_i - x_i|)$$

$$+ 0 \cdot \omega(|x_i - x_{i+1}|) + \ldots + 0 \cdot \omega(|x_i - x_n|) = \omega(0) = 0.$$

Thus Property (ii) is proved.

The continuity of the function $f_0(x)$ for each point $x \in [0, 1]$, different from the nodes $b_1, b_2, \ldots, b_{n-1}$ of the mesh $B$, follows from the continuity of the modulus of continuity $\omega(t)$, from the continuity of the elementary function $y = |x|$ and from the theorem for the continuity of a composition of continuous functions.

Therefore, it remains only to prove that function (2.2.1) is continuous in each of the points $b_1, b_2, \ldots, b_{n-1}$.

Let $b_i$ be an arbitrary fixed node of the mesh $B$ for some $i$ ($1 \leq i \leq n - 1$). Then, according to (1.5.21), (2.2.1) and the continuity of $\omega(t)$ we have

$$\lim_{\substack{x \to b_i \\ x < b_i}} f_0(x) = \lim_{x \to b_i} \omega(|x - x_i|) = \omega(|b_i - x_i|) = \omega\left(\frac{x_{i+1} - x_i}{2}\right)$$

$$\lim_{\substack{x \to b_i \\ x > b_i}} f_0(x) = \lim_{x \to b_i} \omega(|x - x_{i+1}|) = \omega(|b_i - x_{i+1}|) = \omega\left(\frac{x_{i+1} - x_i}{2}\right)$$

Therefore, for each $i = 1, 2, \ldots, n - 1$ the equations

$$\lim_{\substack{x \to b_i \\ x < b_i}} f_0(x) = \lim_{\substack{x \to b_i \\ x > b_i}} f_0(x) = \omega\left(\frac{x_{i+1} - x_i}{2}\right) = f_0(b_i)$$

hold. They allow us to assert that the functions $f_0(x)$ is continuous in each of the points $b_1, b_2, \ldots, b_{n-1}$ also, and hence is continuous over all the segment $[0, 1]$, i.e. $f_0(x) \in C[0, 1]$.

In order to prove that $f_0 \in H^\omega$, it is sufficient to show that for each two points $x$ and $x'$ of the segment $[0, 1]$ the inequality

$$|f_0(x) - f_0(x')| \leqq \omega(|x - x'|) \tag{2.2.5}$$

holds. Let $[b_{s-1}, x_s)$ be those intervals†

$$[b_0, x_1), [x_1, b_1), [b_1, x_2), \ldots, [b_{n-1}, x_m), [x_n, b_n]$$

with the largest length. There exists at least one such interval. According to (1.5.21), along with the interval $[b_{s-1}, x_s)$, the same property has the interval $[x_{s-1}, b_{s-1})$, since

$$x_s - b_{s-1} = b_{s-1} - x_{s-1} = (x_s - x_{s-1})/2.$$

---

† In the case $b_0 = x_1$, $[b_0, x_1)$ is understood to be the empty set.

Therefore, it is not a restriction to consider an interval of the form $[b_{s-1}, x_s)$ only, and not of the form $[x_{s-1}, b_{s-1})$. If there are other intervals with the same extremal property, by $[b_{s-1}, x_s)$ we denote one of them.

Let $x$ and $x'$ be arbitrary points of the interval $[b_{s-1}, x_s)$, From the monotonity and the semi-additivity of the modulus of continuity and from the triangle inequality, we obtain

$$|f_0(x) - f_0(x')| = |\omega(|x - x_s|) - \omega(|x' - x_s|)|$$

$$\leq \omega(||x - x_s| - |x' - x_s||) \leq \omega(|x - x'|)$$

i.e. inequality (2.2.5) is proved for points $x$ and $x'$ of the 'largest' subinterval $[b_{s-1}, x_s)$. Further, let $x$ and $x'$ be arbitrary points of the segment $[0, 1]$. If they both belong to the interval $[b_{s-1}, x_s)$, for those points inequality (2.2.5) is yet to be proved. Therefore, we may assume, that at least one of them belongs to the interval $[b_{s-1}, x_s)$. Let, for example,† $x \in [b_{s-1}, x_s)$ and $x' \in [b_{k-1}, x)$, $k \neq s$. The case when the two points $x$ and $x'$ do not belong to the interval $[b_{s-1}, x_s)$ can be considered in the same way. Since

$$f_0(x') = \omega(|x' - x_k|) = \omega(x_k - x')$$

and

$$0 < x_k - x' \leq x_k - b_{k-1} \leq x_s - b_{s-1}$$

then from the monotonity of $\omega(t)$ the inequalities

$$f_0(x_s) = \omega(0) < f_0(x') = \omega(x_k - x') \leq \omega(x_s - b_{s-1}) = f_0(b_{s-1})$$

follow. Therefore, from the continuity of $f_0(x)$ in the interval $[b_{s-1}, x_s]$ it follows that there is a point $c$ of the interval

$$[b_{s-1}, b_{s-1} - b_{k-1} + x'] \subset [b_{s-1}, x_s)$$

such that

$$f_0(c) = f_0(x') \qquad \text{and} \qquad |x - x'| \geq |x - c|.$$

Using the increasing behaviour of $\omega(t)$, we obtain

$$|f_0(x) - f_0(x')| = |f_0(x) - f_0(c)| \leq \omega(|x - c|) \leq \omega(|x - x'|).$$

Hence, inequality (2.2.5) and Property (iv) are proved.

† The case $x' \in [x_k, b_k)$ is not essentially different.

Then Properties (ii) and (iv) imply (v).

For each function $f \in H_0^\omega$ and for every $x \in [0, 1]$, we obtain

$$f(x) \leqq |f(x)| = \left| \sum_{i=1}^{n} \chi_i(x)[f(x) - f(x_i)] \right| \leqq \sum_{i=1}^{n} \chi_i(x) \cdot |f(x) - f(x_i)|$$

$$\leqq \sum_{i=1}^{n} \chi_i(x) \cdot \omega(|x - x_i|) = f_0(x)$$

i.e. inequality (2.2.3). Then, (2.2.3) and (v) imply (2.2.4) and the lemma is proved.

In figure 2.2.1 the graph of the standard function $f_0(\omega, X; x)$ for a mesh $X$ with three nodes is shown.

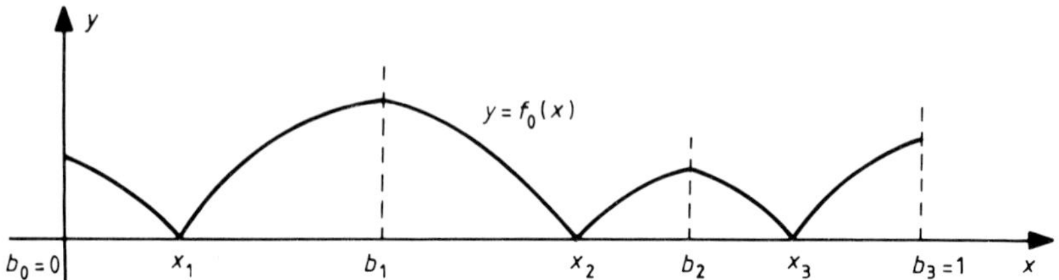


**Figure 2.2.1**   Graph of the standard function $f_0(x)$.

## 2.3   The error of the best method in $H^\omega$

### Lemma 2.3.1

*Let $n \in \mathbb{N}$, $X \in (X)_0$ and $\omega(t)$ ($t \in [0, 1]$) be an arbitrary modulus of continuity. Then, for every fixed $x \in [0, 1]$ the error of the best method for recovery of the functional $Lf = f(x)$ in the class $H^\omega$ for information $T_0(f, X)$ is given by the formula*

$$R(H^\omega, X; x) = f_0(\omega, X; x) \tag{2.3.1}$$

*where $f_0(\omega, X; x)$ is the standard function (2.2.1).*

### Proof

We check the conditions of the Smolyak's lemma. The functionals $Lf = f(x)$, $L_i f = f(x_i)$ ($i = 1, 2, \ldots, n$) are linear. They are defined in the normed linear space $C[0,1]$. According to Theorem A (Korneychuk), the set $H^\omega$ is convex, centrally symmetric with the zero point as the centrum body in $C[0,1]$. By Lemma 2.2.1 (see (2.2.4)), (2.2.1) and (2.2.2), we have

$$\sup\{Lf : f \in H_0^\omega\} = f_0(x) = \sum_{i=1}^{n} \chi_i(x)\omega(|x - x_i|) \leqq \omega(1). \qquad (2.3.2)$$

Therefore, all the conditions of Smolyak's lemma are satisfied. According to the corollary of this lemma, the error $R(H^\omega, X; x)$ of the best method for recovery of the functional $Lf = f(x)$ in $H^\omega$ on $T_0(f; X)$ is given by

$$R(H^\omega, X; x) = \sup\{Lf = f(x) : f \in H_0^\omega\}.$$

From this equality and from (2.3.2) we obtain (2.3.1), and thus the lemma is proved.

## Theorem 2.3.1

*Let $n \in \mathbb{N}$, $X \in (X)_0$, $\omega(t)$ be an arbitrary modulus of continuity, and $x \in [0,1]$ be a fixed point of the segment $[0,1]$.*

*Out of all feasible methods for recovery of the functional $Lf = f(x)$ in the class $H^\omega$ on the information $T_0(f; X)$, the best is*

$$\Phi^0(x) = \Phi^0(f, X; x) = \sum_{i=1}^{n} \chi_i(x) \cdot f(x_i), \qquad x \in [0,1] \qquad (2.3.3)$$

*constructed by formula (1.5.25) for $r = 0$.*

*For each function $f \in H^\omega$ and for every fixed $x \in [0,1]$*

$$|f(x) - \Phi^0(f, X; x)| \leqq f_0(\omega, X; x) \qquad (2.3.4)$$

*where $f_0(\omega, X; x)$ is the standard function (2.2.1).*

*The equality sign in (2.3.4) is attained by $f_0(\omega, X; x)$.*

## Proof

For every fixed $x \in [0,1]$ and for each function $f \in H^\omega$ relations (2.3.3), (1.5.24) and (2.2.1) imply

$$|f(x) - \Phi^0(f, X; x)| = \left| \sum_{i=1}^{n} \chi_i(x)[f(x) - f(x_i)] \right|$$

$$\leq \sum_{i=1}^{n} \chi_i(x)|f(x) - f(x_i)| \leq \sum_{i=1}^{n} \chi_i(x)\cdot\omega(|x - x_i|)$$

$$= f_0(\omega, X; x)$$

and thus inequality (2.3.4) is proved.

Since according to Lemma 2.2.1 $f_0(\omega, X; x) \geq 0$, $f_0 \in H_0^\omega$, then from (2.3.3), for each $x \in [0,1]$, the equality $\Phi^0(f_0, X; x) = 0$ follows, and with it, the equation

$$|f_0(\omega, X; x) - \Phi^0(f_0, X; x)| = f_0(\omega, X; x). \tag{2.3.5}$$

From (2.3.5), (2.3.4) and $f_0 \in H^\omega$ (see Lemma 2.2.1) we get

$$\sup\{|f(x) - \Phi^0(f, X; x)| : f \in H^\omega\} = f_0(\omega, X; x), \qquad x \in [0,1]. \tag{2.3.6}$$

Now, (2.3.6) and (2.3.1) imply

$$R(H^\omega, X; x) = \sup_{f \in H^\omega} |f(x) - \Phi^0(f, X; x)| = R(H^\omega, X, \Phi^0; x). \tag{2.3.7}$$

According to Definition (1.5.3), this equation is sufficient in order to assert that method (2.3.3) is the best one for the recovery of the functional $Lf = f(x)$ in $H^\omega$ on $T_0(f, X)$.

From (2.3.5) it follows also that the equality sign in (2.3.4) is attained by the standard function (2.2.1).

The proof is complete.

**Corollary 2.3.1**   *Let $n \in \mathbb{N}$, $X \in (X)_0$ and $\omega(t), t \in [0,1]$, be an arbitrary modulus of continuity. Then the error of the best method for recovery of functions of the class $H^\omega$ on the information $T_0(f; X)$ in the uniform and integral $(L_p, p > 0)$ metrics is given by*

$$R(H^\omega, X)_C = \| f_0(\omega, X; .)\|_C = \omega(D(B, X)_\infty) \tag{2.3.8}$$

*and*

$$R(H^\omega, X)_{L_p} = \| f_0(\omega, X; .)\|_{L_p} = \left[ \sum_{i=1}^{n} \int_{b_{i-1}}^{b_i} \omega^p(|x - x_i|)dx \right]^{1/p}, \quad p > 0 \tag{2.3.9}$$

*respectively.*

## **Proof**

From (1.5.12), (2.3.1), (2.2.1), (1.6.13), (1.5.21) and from the monotonicity and continuity of $\omega(t)$ it follows that

$$
R(H^\omega, X)_C = \|R(H^\omega, X; .)\|_C = f_0(\omega, X; .)\|_C
$$

$$
= \left\| \sum_{i=1}^{n} \chi_i(x) \cdot \omega(|x - x_i|) \right\|_C
$$

$$
= \max\{ \max_{b_{i-1} \le x \le b_i} \omega(|x - x_i|) : i = 1, 2, \ldots, n \}
$$

$$
= \max\{\omega(|b_i - x_i|, \omega(|b_{i-1} - x_i|) : i = 1, 2, \ldots, n\}
$$

$$
= \omega(\max\{|b_i - x_i|, |b_{i-1} - x_i| : i = 1, 2, \ldots, n\})
$$

$$
= \omega(D(B, X)_\infty)
$$

and thus (2.3.8) is proved.

From (1.5.13), (2.3.1) and (2.2.1) we have

$$
R(H^\omega, X)_{L_p} = \|R(H^\omega, X; .)\|_{L_p} = \|f_0(\omega, X; .)\|_{L_p}
$$

$$
= \left\{ \int_0^1 [f_0(\omega, X; x)]^p \, dx \right\}^{1/p}
$$

$$
= \left\{ \sum_{i=1}^{n} \int_{b_{i-1}}^{b_i} [f_0(\omega, X; x)]^p \, dx \right\}^{1/p}
$$

$$
= \left\{ \sum_{i=1}^{n} \int_{b_{i-1}}^{b_i} \omega^p(|x - x_i|) \, dx \right\}^{1/p}, \qquad p > 0.
$$

**Corollary 2.3.2**   *Let $n \in \mathbb{N}$, $X \in (X)_0$ and $\omega(t), 0 \le t \le 1$, be an arbitrary modulus of continuity. Then the following extremal relations hold*

$$
R(H^\omega, X)_C = \sup_{f \in H^\omega} \|f(.) - \Phi^0(f, X; .)\|_C \tag{2.3.10}
$$

$$
R(H^\omega, X)_{L_p} = \sup_{f \in H^\omega} \|f(.) - \Phi^0(f, X; .)\|_{L_p}, \qquad p > 0. \tag{2.3.11}
$$

## Proof

Indeed, according to (1.5.12) and (2.3.7) we have

$$R(H^\omega, X)_C = \sup_{0 \le x \le 1} R(H^\omega, X; x) = \sup_{0 \le x \le 1} \sup_{f \in H^\omega} |f(x) - \Phi^0(f, X; x)|. \qquad (2.3.12)$$

Having in mind that the supremums can be interchanged (see [6], p 304, formula (D.17)), we obtain the equation

$$\sup_{0 \le x \le 1} \sup_{f \in H^\omega} |f(x) - \Phi^0(f, X; x)| = \sup_{f \in H^\omega} \|f(.) - \Phi^0(f, X; .)\|_C$$

which, along with (2.3.12) gives (2.3.10).

Using (2.3.4) and the monotonity of the norm $\|.\|_p$ ($p > 0$) (see e.g. [21], p 279) then for each function $f \in H^\omega$ we obtain

$$\|f(.) - \Phi^0(f, X; .)\|_{L_p} \le \|f_0(\omega, X; .)\|_{L_p}, \qquad p > 0.$$

From (2.3.5) it follows correspondingly that

$$\|f_0(\omega, X; .) - \Phi^0(f_0, X; .)\|_{L_p} = \|f_0(\omega, X; .)\|_{L_p}, \qquad p > 0.$$

Since $f_0(\omega, X; x) \in H^\omega$, then the last two relations along with (2.3.9) imply (2.3.11). Thus the assertion of the corollary is proved.

## 2.4　Optimal methods in $H^\omega$

### Theorem 2.4.1

*Let $n \in \mathbb{N}$, $v$ be a fixed element of the set $\{0, 1, 2, 3\}$, let the mesh† $X$ belong to the set $(X)_v$, and $\omega(t)$ ($t \in [0, 1]$) be an arbitrary modulus of continuity.*

*Then, out of all methods for recovery of functions of the class $H^\omega$ on the information $T_0(f; X)$, built for a mesh $X \in (X)_v$, the method $\Phi^0(f, X_v; x)$ formed according to formula (2.3.3) for the mesh $X_v$, defined in $(1.6.5_v)$ is the optimal one both in uniform and in integral $(L_p, p > 0)$ metrics.*

*For the error in $C$ and $L_p$ ($p > 0$) the following formulas hold*

$$R_n^v(H^\omega)_C = \omega(D(A_v, X_v)_\infty) \qquad (2.4.1)$$

---

† See the notation of Section 1.6.

*and*

$$R_n^v(H^\omega)_{L_p} = \left\{ \int\limits_0^1 \omega^p(u \cdot D(A_v, X_v)_\infty) du \right\}^{1/p}, \qquad p > 0 \qquad (2.4.2)$$

*where $D(A, X)_p$ $(0 < p \leq \infty)$ is the deviation of the mesh pair $(A, X)$ and the mesh pair $(A_v, X_v)$ is defined in $(1.6.5_v)$. The standard function $f_0(\omega, X; x)$, defined for the mesh $X_v$ by $(2.2.1)$ is extremal both in $L_p$ $(p > 0)$ metrics.*

## Proof

From Definition (1.5.15) and from Corollary 2.3.1 it follows that

$$\begin{aligned}
R_n^v(H^\omega)_C &= \inf\{R(H^\omega, X)_C : X \in (X)_v\} \\
&= \inf\{\omega(D(B, X)_\infty) : x \in (X)_v\} \\
&= \inf\{\omega(D(B, X)_\infty) : (B, X) \in (B, X)_v\} \\
&\geq \inf\{\omega(D(A, X)_\infty) : (A, X) \in (A, X)_v\}. \qquad (2.4.3)
\end{aligned}$$

For the last inequality we used the relation

$$(B, X)_v \subset (A, X)_v$$

which holds due to the unicity of the mesh $B$, determined by $X$ according to formulas (1.5.21) which is not certain for the mesh $A$.

Here and further on we use the following lemma.

## Lemma 2.4.1

*Let $P$ be an arbitrary non-empty set, $\mathbb{R}_+ = \{x : x \in \mathbb{R}, x \geq 0\}$, $\varphi : P \to \mathbb{R}_+$, and $\omega(t)$, $0 \leq t < \infty$, be an arbitrary modulus of continuity. Then*

$$\inf\{\omega(\varphi(x) : x \in P\} = \omega(\inf\{\varphi(x) : x \in P\}). \qquad (2.4.4)$$

---

**Remark**   Using this assertion, we say that the modulus of continuity and the infimum sign are interchangeable.

---

## **Proof**

From $\varphi: P \to \mathbb{R}_+$ it follows that for each element $x \in P$ the real number $\varphi(x)$ is non-negative. Since the set $\{\varphi(x): x \in P\}$ of real numbers is bounded from below, then there exists the infimum

$$\inf\{(\varphi(x): x \in P\} = \alpha \tag{2.4.5}$$

that is

(i) $\forall x \in P: \alpha \leq \varphi(x)$
(ii) $\forall \delta > 0, \exists x_0 \in P: \varphi(x_0) < \alpha + \delta$.
Further, let $\varepsilon > 0$ be an arbitrary positive number. Then, let us choose a positive number $\delta_0$ such that for each $\delta \in (0, \delta_0)$ the inequality

$$\omega(\delta) < \varepsilon \tag{2.4.6}$$

is satisfied. Such a choice is possible, since

$$\lim_{\substack{\delta \to 0 \\ \delta > 0}} \omega(\delta) = 0.$$

Let $\delta_1$ is a fixed positive number, smaller than $\delta_0$. According to (2.4.6), we have

$$\omega(\delta_1) < \varepsilon \tag{2.4.7}$$

but, according to (ii) there is a $x_1 \in P$ such that the inequality

$$\varphi(x_1) < \alpha + \delta_1 \tag{2.4.8}$$

The increasing behaviour of $\omega(t)$, (i) and (2.4.8) imply the inequalities
(i') $\forall x \in P: \omega(\alpha) \leq \omega(\varphi(x))$
(ii') $\exists x_1 \in P: \omega(\varphi(x_1)) \leq \omega(\alpha + \delta_1)$.
Then, using the semi-additivity of $\omega(t)$ and inequality (2.4.7), from (ii') we obtain that: for every $\varepsilon > 0$ there exists $x_1 \in P$ such that the inequality

$$\omega(\varphi(x_1)) \leq \omega(\alpha + \delta_1) \leq \omega(\alpha) + \omega(\delta_1) < \omega(\alpha) + \varepsilon$$

holds.

From this inequality, from (i') and the definition of the infimum, we have

$$\inf\{\omega(\varphi(x)): x \in P\} = \omega(\alpha) \overset{(2.4.5)}{=} \omega(\inf\{\varphi(x): x \in P\}).$$

The lemma is proved.

Using Lemma 2.4.1 and Corollary 2.1.3, we obtain

$$\inf\{\omega(D(A,X)_\infty:(A,X)\in(A,X)_v\} = \omega(\inf\{D(A,X)_\infty:(A,X)\in(A,X)_v\})$$

$$= \omega(D(A_v,X_v)_\infty).$$

Applying again Corollary 2.3.1, from the last relation and (2.4.3), we get the inequality

$$R_n^v(H^\omega)_C = \inf\{R(H^\omega,X)_C:X\in(X)_v\} \geq \omega(D(A_v,X_v)_\infty) = R(H^\omega,X_v)_C. \quad (2.4.9)$$

Indeed, according to Definition $(1.6.5_v)$ of the mesh pair it is immediate to verify that for a fixed $v\in\{0,1,2,3\}$ the mesh $A_v$ can be obtained from the mesh $X_v$ by the formulas (1.5.21). Taking into account that $X_v\in(X)_v$, then from (2.4.9) we obtain the equality

$$R_n^v(H^\omega)_C = R(H^\omega,X_v)_C = \omega(D(A_v,X_v)_\infty). \quad (2.4.10)$$

This equality implies:
(1) equality (2.4.1);
(2) the optimality of the mesh $X_v$ in the uniform metrics and
(3) (due to $A_v = B_v$) the optimality of the method $\Phi^0(f,X_v;x)$, built according to formula (3.3) by the mesh $X_v$ in the class $H^\omega$ and in the uniform metrics.
Finally, the chain of equalities

$$R_n^v(H^\omega)_C \overset{(2.4.10)}{=} R(H^\omega,X_v)_C \overset{(2.3.8)}{=} \|f_0(\omega,X_v;.)\|_C$$

$$\overset{(2.3.5)}{=} \|f_0(\omega,X_v;.) - \Phi^0(f_0,X_v;.)\|_C$$

proves the extremality of the standard function $f_0(\omega,X_v;x)$ in the uniform metrics. It remains only to prove the assertions of the theorem for integral $(L_p, p>0)$ metrics.
By Definition (1.5.15) we have

$$R_n^v(H^\omega)_{L_p} = \inf\{R(H^\omega,X)_{L_p}:X\in(X)_v\}. \quad (2.4.11)$$

Since the function $\varphi(u) = \omega^p(u) = [\omega(u)]^p$, $u\in[0,1]$ is increasing, then by the optimization theorem and Corollary 2.3.1, for each mesh $X\in(X)_v$ we obtain

$$R^p(H^\omega,X)_{L_p} \overset{(2.3.9)}{=} [R(H^\omega,X)_{L_p}]^p = \sum_{i=1}^n \int_{b_{i-1}}^{b_i} \omega^p(|x-x_i|)dx$$

$$\overset{(2.1.2)}{=} L_n(\omega^p;B,X) \overset{(2.1.1)}{\geq} L_n(\omega^p;A_v,X_v) \overset{(2.3.9)}{=} R^p(H^\omega,X_v)_{L_p}$$

In the last equality we have used again that $A_v = B_v$, therefore

$$R(H^\omega, X)_{L_p} \geq R(H^\omega, X_v)_{L_p} = [L_n(\omega^p; A_v, X_v)]^{1/p}, \qquad p > 0.$$

Taking into account that $X_v \in (X)_v$, then from the last inequality it follows that

$$\inf\{R(H^\omega, X)_{L_p} : X \in (X)^v\} = [L_n(\omega^p; A_v, X_v)]^{1/p}. \tag{2.4.12}$$

Now, from (2.4.11), (2.4.12) and (2.1.3), we get

$$R_n^v(H^\omega)_{L_p} = R(H^\omega, X_v)_{L_p} = \left\{ \int_0^1 \omega^p(u \cdot D(A_v, X_v)_\infty) du \right\}^{1/p}, \qquad p > 0. \tag{2.4.13}$$

The equality (2.4.13) implies:
(1) equality (2.4.2);
(2) the optimality of the mesh $X_v$ in $L_p$ metrics $(p > 0)$ also and
(3) (due to $A_v = B_v$) the optimality of the method $\Phi^0(f, X_v; x)$, built according to formula (2.3.3) from the mesh $X_v$ in the class $H^\omega$ and in the integral metrics.

Finally, the chain of equalities

$$R_n^v(H^\omega)_{L_p} \overset{(2.4.13)}{=} R(H^\omega, X_v)_{L_p} \overset{(2.3.9)}{=} \|f_0(\omega, X_v; .)\|_{L_p}$$

$$\overset{(2.3.5)}{=} \|f_0(\omega, X_v; .) - \Phi^0(f_0, X_v; .)\|_{L_p}, \qquad p > 0$$

proves the extremality of the standard function $f_0(\omega, X_v; x)$ in the integral metrics also.

Thus the theorem is completely proved.


Now we describe explicitly the errors of the optimal methods in $L_p$ $(p > 0)$ metrics for various meshes (with, or without restriction):

$$R_n^0(H^\omega)_{L_p} = \left\{ \int_0^1 \omega^p(u/2n) du \right\}^{1/p} \tag{2.4.14$_0$}$$

$$R_n^1(H^\omega)_{L_p} = \left\{ \int_0^1 \omega^p(u/(2n-1)) du \right\}^{1/p} \tag{2.4.14$_1$}$$

$$R_n^2(H^\omega)_{L_p} = \left\{ \int_0^1 \omega^p(u/(2n-2))\mathrm{d}u \right\}^{1/p} \qquad (2.4.14_2)$$

$$R_n^3(H^\omega)_{L_p} = \left\{ \int_0^1 \omega^p(u/(2n-1))\mathrm{d}u \right\}^{1/p} \qquad (2.4.14_3)$$

Especially, for $p = \infty$ in uniform metrics for the same meshes, we have

$$R_n^0(H^\omega)_C = \omega(1/2n) \qquad (2.4.15_0)$$

$$R_n^1(H^\omega)_C = \omega(1/(2n-1)) \qquad (2.4.15_1)$$

$$R_n^2(H^\omega)_C = \omega(1/(2n-2)) \qquad (2.4.15_2)$$

$$R_n^3(H^\omega)_C = \omega(1/(2n-1)). \qquad (2.4.15_3)$$

Further, let us write down the extremal relations, following from (2.4.10) and (2.4.13), and from (2.3.10) and (2.3.11), respectively:

$$R_n^\nu(H^\omega)_C = \sup_{f \in H^\omega} \| f(.) - \Phi^0(f, X_\nu; .) \|_C \qquad (2.4.16)$$

and

$$R_n^\nu(H^\omega)_{L_p} = \sup_{f \in H^\omega} \| f(.) - \Phi^0(f, X_\nu; .) \|_{L_p}, \qquad p > 0 \qquad (2.4.17)$$

for each fixed $\nu \in \{0, 1, 2, 3\}$.

## 2.5  Recovery of functions of the classes $W^r H^\omega$

Here we are looking for optimal methods of the form (1.5.25) for recovery of functions of the classes $W^r H^\omega$ in uniform and in integral metrics. In other words, we are solving the extremal problem (1.5.29), (1.5.30) in a restricted setting.

Let $n \in \mathbb{N}$, $r \in \mathbb{N}$, $X \in (X)_0$, $\omega(t)$ $(t \in [0,1])$ be an arbitrary modulus of continuity and $B = (b_0, b_1, \ldots, b_n)$ be the mesh, which is uniquely determined by the mesh $X$ by means of formulas (1.5.21).

For fixed $\alpha$ and $x$ from the segment $[0,1]$ and for each function $f \in C^r$, let us denote

$$g(z) = f(\alpha + z(x - \alpha)), \qquad z \in [0, 1].$$

From Taylor's formula with the remainder term in integral form

$$g(1) = \sum_{k=0}^{r} \frac{g^{(k)}(0)}{k!} + \frac{1}{(r-1)!} \int_0^1 (1-z)^{r-1}[g^{(r)}(z) - g^{(r)}(0)]dz$$

and from

$$g^{(k)}(z) = f^{(k)}(\alpha + z(x - \alpha)) \cdot (x - \alpha)^k, \qquad k = 0, 1, \ldots, r,$$

we get the formula

$$f(x) = \sum_{k=0}^{r} \frac{f^{(k)}(\alpha)}{k!} (x - \alpha)^k + \frac{(x-\alpha)^r}{(r-1)!}$$

$$\times \int_0^1 (1-z)^{r-1}[f^{(r)}(\alpha + z(x - \alpha)) - f^{(r)}(\alpha)]dz, \quad x \in [0, 1].$$

$$(2.5.1)$$

### Lemma 2.5.1

*Let $n \in \mathbb{N}$, $r \in \mathbb{N}$, $X \in (X)_0$ and $\omega(t)$ $(t \in [0, 1])$ be an arbitrary modulus of continuity. The function*

$$f_r(x) \equiv f_r(\omega, X; x) = \frac{1}{(r-1)!} \int_0^x (x - t)^{r-1} f_0(\omega, X; t)dt, \quad x \in [0, 1] \quad (2.5.2)$$

*where $f_0(\omega, X; x)$, $x \in [0, 1]$ is the standard function (1.2.1), has the following properties:*
 (i) $f_r(\omega, X; x) \geq 0$, $x \in [0, 1]$;
 (ii) $f_r^{(k)}(\omega, X; 0) = 0$ $(k = 0, 1, 2, \ldots, r-1)$;
 (iii) $f_r(\omega, X; x) \in C^r$;
 (iv) $f_r^{(k)}(\omega, X; x) = f_{r-k}(\omega, X; x)$, $k = 0, 1, \ldots; r$ $(x \in [0, 1])$;
 (v) $f_r(\omega, X; x) \in W^r H^\omega$;
 (vi) $f_r(\omega, X; x) \in W^r H_0^\omega = \{f(x): f \in W^r H^\omega, \quad f^{(r)}(x_i) = 0, \quad i = 1, 2, \ldots, n\}$;

*(vii) for each $x \in [0, 1]$ and for even $r \in \mathbb{N}$ it holds that*

$$f_r(\omega, X; x) - \Phi^r(f_r, X; x) = \sum_{i=1}^{n} \chi_i(x) \frac{|x - x_i|^r}{(r-1)!} \cdot \int_0^1 (1 - z)^{r-1} \omega(z|x - x_i|)\,dz$$

$$(2.5.3)$$

*where the quasi-spline $\Phi^r(f, X; x)$ is defined in (1.5.25) and the characteristic function $\chi_i(x)$ in (1.5.22).*

### Proof

Since (i)–(vi) follow directly from (2.5.2) and the properties of the standard function (2.2.1), we shall prove only property (vii).

From the definition of the quasi-spline $\Phi^r(f_r, X; x)$ (see (1.5.15)), along with (2.5.1), (2.5.2) and the properties of the standard functions (2.2.1) (see Lemma 2.2.1), for even $r$ we obtain successively

$$f_r(\omega, X; x) - \Phi^r(f_r, X; x) = \sum_{i=1}^{n} \chi_i(x)[f_r(\omega, X; x) - S_r(f_r, x_i; x)]$$

$$\overset{(2.5.1)}{=} \sum_{i=1}^{n} \chi_i(x) \frac{(x - x_i)^r}{(r-1)!} \int_0^1 (1 - z)^{r-1}[f_r^{(r)}(x_i + z(x - x_i)) - f_r^{(r)}(x_i)]\,dz$$

$$= \sum_{i=1}^{n} \chi_i(x) \frac{|x - x_i|^r}{(r-1)!} \int_0^1 (1 - z)^{r-1}[f_0(x_i + z(x - x_i)) - f_0(x_i)]\,dz$$

$$= \sum_{i=1}^{n} \chi_i(x) \frac{|x - x_i|^r}{(r-1)!} \int_0^1 (1 - z)^{r-1} \cdot \omega(z|x - x_i|)\,dz$$

and the proof is completed.

### Theorem 2.5.1

*Let $n \in \mathbb{N}$, $X \in (X)_0$, $r(r \in \mathbb{N})$ be an even number, and $\omega(t)$ $(t \in [0, 1])$ be an arbitrary modulus of continuity. Then for each $x \in [0, 1]$ it holds that*

$$\Delta(W^r H^\omega, X; x) = \sup_{f \in W^r H^\omega} |f(x) - \Phi^r(f, X; x)|$$

$$= \sum_{i=1}^{n} \chi_i(x) \cdot \frac{|x - x_i|^r}{(r-1)!} \cdot \int_0^1 (1-z)^{r-1} \cdot \omega(z|x - x_i|) \mathrm{d}z \quad (2.5.4)$$

*where the quasi-spline $\Phi^r(f, X; x)$ is defined in (1.5.25) and the characteristic function $\chi_i(x)$ in (1.5.22). The supremum in (2.5.4) is attained by the function (2.5.2).*

## Proof

For every function $f \in W^r H^\omega$ $(r = 2, 4, \ldots)$ and for each $x \in [0, 1]$, by (1.5.25) and (2.5.1) we have

$$|f(x) - \Phi^r(f, X; x)|$$

$$= \left| \sum_{i=1}^{n} \chi_i(x)[f(x) - S_r(f, x_i; x)] \right|$$

$$\leq \sum_{i=1}^{n} \chi_i(x)|f(x) - S_r(f, x_i; x)|$$

$$= \sum_{i=1}^{n} \chi_i(x) \cdot \left| \frac{(x - x_i)^r}{(r-1)!} \int_0^1 (1-z)^{r-1}[f^{(r)}(x_i + z(x - x_i)) - f^{(r)}(x_i)] \mathrm{d}z \right|$$

$$\leq \sum_{i=1}^{n} \chi_i(x) \frac{|x - x_i|^r}{(r-1)!} \cdot \int_0^1 (1-z)^{r-1} \omega(z|x - x_i|) \mathrm{d}z \quad (2.5.5)$$

that is

$$\Delta(W^r H^\omega, X; x) = \sup_{f \in W^r H^\omega} |f(x) - \Phi^r(f, X; x)|$$

$$\leq \sum_{i=1}^{n} \chi_i(x) \cdot \frac{|x - x_i|^r}{(r-1)!} \cdot \int_0^1 (1-z)^{r-1} \cdot \omega(z|x - x_i|) \mathrm{d}z \quad (2.5.5')$$

Using (2.5.5′) and (2.5.3) and taking into account that $f_r(\omega, X; x)$ belongs to the class $W^r H^\omega$, we obtain (2.5.4).

From (2.5.3) it follows that the supremum in (2.5.4) is attained by the function (2.5.2) and thus the proof is completed.

**Corollary 2.5.1** *Let $n \in \mathbb{N}$, $X \in (X)_0$, $r (r \in \mathbb{N})$ be an even number, and $\omega(t)$ $(t \in [0, 1])$ be an arbitrary modulus of continuity. Then*

$$\Delta(W^r H^\omega; X)_C = \sup_{f \in W^r H^\omega} \| f(.) - \Phi^r(f, X; .) \|_C$$

$$= \frac{D^r(B, X)_\infty}{(r-1)!} \cdot \int_0^1 (1-z)^{r-1} \cdot \omega(z \cdot D(B, X)_\infty) dz \qquad (2.5.6)$$

*and*

$$\Delta(W^r H^\omega, X)_{L_p} = \sup_{f \in W^r H^\omega} \| f(.) - \Phi^r(f, X; .) \|_{L_p}$$

$$= \left\{ \sum_{i=1}^n \int_{b_{i-1}}^{b_i} \left[ \frac{|x - x_i|^r}{(r-1)!} \int_0^1 (1-z)^{r-1} \omega(z|x - x_i|) dz \right]^p dx \right\}^{1/p},$$

$$p > 0 \qquad (2.5.7)$$

*where the quasi-spline $\Phi^r(f, X; x)$ is determined by (1.5.25), the mesh $B = (b_0, b_1, \ldots, b_n)$ by (1.5.21) and $D(A, X)_\infty$ is the uniform deviation of the mesh pair $(A, X)$.*

### Proof

First we prove (2.5.6). From (2.5.4) and from the definition of $\Delta(W^r H^\omega, X)_C$ it follows that

$$\Delta(W^r H^\omega, X)_C = \sup_{0 \leq x \leq 1} \sup_{f \in W^r H^\omega} |f(x) - \Phi^r(f, X; x)|. \qquad (2.5.8)$$

Since the supremums are interchangeable (see [6], p 304, formula (D17)), then

$$\sup_{0 \leq x \leq 1} \sup_{f \in W^r H^\omega} |f(x) - \Phi^r(f, X; x)| = \sup_{f \in W^r H^\omega} \| f(.) - \Phi^r(f, X; .) \|_C. \qquad (2.5.9)$$

From (2.5.8) and (2.5.9) the first of the equations, (2.5.6), follows.
Let us denote

$$\varphi(u) = \frac{u^r}{(r-1)!} \int_0^1 (1-z)^{r-1} \cdot \omega(z \cdot u) dz, \quad u \in [0,1]. \tag{2.5.10}$$

Function (2.5.10) is continuous and increasing in the segment $[0,1]$. Again from the definition of $\Delta(W^r H^\omega, X)_C$ and (2.5.4) we get

$$\Delta(W^r H^\omega, X)_C = \sup_{0 \le x \le 1} \Delta(W^r H^\omega, X; x)$$

$$\overset{(2.5.4)}{=} \sup_{0 \le x \le 1} \sum_{i=1}^{n} \chi_i(x) \frac{|x-x_i|^r}{(r-1)!} \int_0^1 (1-z)^{r-1} \omega(z|x-x_i|dx$$

$$\overset{(2.5.10)}{=} \sup_{0 \le x \le 1} \sum_{i=1}^{n} \chi_i(x) \cdot \varphi(|x-x_i|)$$

$$\overset{\varphi \in C}{=} \max_{1 \le i \le n} \sup_{b_{i-1} \le x \le b_i} \varphi(|x-x_i|)$$

$$\overset{\varphi \uparrow}{=} \max\{\varphi(|b_{i-1}-x_i|), \varphi(|b_i-x_i|): i = 1,2,\ldots,n\}$$

$$\overset{\varphi \uparrow}{=} \varphi(\max\{|b_{i-1}-x_i|, |b_i-x_i|: i = 1,2,\ldots,n\})$$

$$\overset{(1.6.13)}{=} \varphi(D(B,X)_\infty)$$

$$\overset{(2.5.10)}{=} \frac{D^r(B,X)_\infty}{(r-1)!} \int_0^1 (1-z)^{r-1} \cdot \omega(z \cdot D(B,X)_\infty) dz.$$

Thus the second equation of (2.5.6) is proved.

Now, let us prove (2.5.7). On the one hand, from the definition of $\Delta(W^r H^\omega, X)_{L_p}$ $(p > 0)$, (2.5.4) and (2.5.10) it follows immediately that

$$\Delta(W^r H^\omega, X)_{L_p} = \|\Delta(W^r H^\omega, X; .)\|_{L_p}$$

$$(2.5.4) = \left\{ \sum_{i=1}^{n} \int_{b_{i-1}}^{b_i} \left[ \frac{|x - x_i|^r}{(r-1)!} \int_0^1 (1-z)^{r-1} \cdot \omega(z|x - x_i|) dz \right]^p dz \right\}^{1/p}$$

$$(2.5.10) = \left\{ \sum_{i=1}^{n} \int_{b_{i-1}}^{b_i} \varphi^p(|x - x_i|) dx \right\}^{1/p}, \quad p > 0 \qquad (2.5.11)$$

and thus the second of equalities (2.5.7) is obtained.

On the other hand, from (2.5.5) and from the monotonity of the function $y = x^p$ $(x \geq 0, (>0)$ for each function $f \in W^r H^\omega$ the inequality

$$\|f(.) - \Phi^r(f, X; .)\|_{L_p} \leq \left\{ \sum_{i=1}^{n} \int_{b_{i-1}}^{b_i} \varphi^p(|x - x_i|) dx \right\}^{1/p}, \quad p > 0, \quad (2.5.12)$$

holds.

From (2.5.3) it follows that

$$\|f_r(.) - \Phi^r(f_r, X; .)\|_{L_p} = \left\{ \sum_{i=1}^{n} \int_{b_{i-1}}^{b_i} \varphi^p(|x - x_i|) dx \right\}^{1/p}, \quad p > 0. \quad (2.5.13)$$

Keeping in mind that $f_r(\omega, X; x) \in W^r H^\omega$, then from (2.5.12) and (2.5.13) we get the equation

$$\sup_{f \in W^r H^\omega} \|f(.) - \Phi^r(f, X; .)\|_{L_p} = \left\{ \sum_{i=1}^{n} \int_{b_{i-1}}^{b_i} \varphi^p(|x - x_i|) dx \right\}^{1/p}, \quad p > 0. \quad (2.5.14)$$

From (2.5.14) and (2.5.11) the first equation of (2.5.7) follows. The proof is completed.

## Theorem 2.5.2

*Let $n \in \mathbb{N}$, $r(r \in \mathbb{N})$ be an even number, $v$ be an arbitrary element of the set $\{0, 1, 2, 3\}$, $X \in (X)_v$ and $\omega(t)$ $(t \in [0, 1])$ be an arbitrary modulus of continuity. Then out of all methods of the form (1.5.25) for recovery of functions of the class $W^r H^\omega$, the method $\Phi^r(f, X_v; x)$ built on formula (1.5.25) by means of the mesh $X_v$, is optimal both in uniform and in integral $(L_p, p > 0)$ metrics.*

*Furthermore,*

$$\Delta_n^v(W^r H^\omega)_C = \inf_{X \in (X)_v} \Delta(W^r H^\omega, X)_C = \Delta(W^r H^\omega, X_v)_C$$

$$= \sup_{f \in W^r H^\omega} \| f(.) - \Phi^r(f, X_v; .) \|_C$$

$$= \frac{D^r(A_v, X_v)_\infty}{(r-1)!} \int_0^1 (1-z)^{r-1} \omega(z \cdot D(A_v, X_v)_\infty) dz \quad (2.5.15)$$

*and*

$$\Delta_n^v(W^r H^\omega)_{L_p} = \inf_{X \in (X)_v} \Delta(W^r H^\omega, X)_{L_p} = \Delta(W^r H^\omega, X_v)_{L_p}$$

$$= \sup_{f \in W^r H^\omega} \| f(.) - \Phi^r(f, X_v; .) \|_{L_p}$$

$$= \frac{D^r(A_v, X_v)_\infty}{(r-1)!} \left\{ \int_0^1 \left[ \int_0^1 u^r (1-z)^{r-1} \omega(zuD(A_v, X_v)_\infty) dz \right]^p du \right\}^{1/p}.$$

$$(2.5.16)$$

*The supremums in (2.5.15) and (2.5.16) are attained by the function $f_r(\omega, X_v; x)$ defined by (2.5.2).*

## Proof

First, we prove (2.5.15). By the definition of $\Delta_n^v(W^r H^\omega)_C$, Corollary 2.5.1 and (2.5.10) it follows that

$$\Delta_n^v(W^r H^\omega)_C = \inf_{X \in (X)_v} \Delta(W^r H^\omega; X)_C \overset{\substack{(2.5.6)\\(2.5.10)}}{=} \inf_{X \in (X)^v} \varphi(D(B, X)_\infty).$$

Using the increasing behaviour of function (2.5.10), from (2.1.19) we get

$$\inf_{X \in (X)_v} \varphi(D(B, X)_\infty) = \inf_{(B,X) \in (A,X)_v} \varphi(D(B, X)_\infty) \geq \inf_{(A,X) \in (A,X)_v} \varphi(D(A, X)_\infty)$$

$$= \varphi(D(A_v, X_v)_\infty) = \varphi(D(B_v, X_v)_\infty) \overset{(2.5.6)}{=} \Delta(W^r H^\omega, X_v)_C$$

Finally, noting that $X_v \in (X)_v$, we find that

$$\Delta_n^v(W^r H^\omega)_C = \Delta(W^r H^\omega, X_v)_C = \varphi(D(A_v, X_v)_\infty)$$

$$\overset{(2.5.6)}{=} \sup_{f \in W^r H^\omega} \|f(.) - \Phi^r(f, X; .)\|_C.$$

Therefore, relations (2.5.15) are proved. From (2.5.15) follows the optimality of the mesh $X_v$ and the method $\Phi^r(f, X_v; x)$ with respect to all the methods of the form (1.5.25) built on meshes $x \in (X)_v$ for recovery of functions of the classes $W^r H^\omega$ $(r = 2, 4, 6, \ldots)$ in uniform metrics.

By Theorem 2.5.1 and from the interchangeability of the supremums, we get

$$\|f_r(.) - \Phi^r(f_r, X_v; .)\|_C = \sup_{0 \le x \le 1} \sup_{f \in W^r H^\omega} |f(x) - \Phi^r(f, X_v; x)|$$

$$= \sup_{f \in W^r H^\omega} \|f(.) - \Phi^r(f, X_v; .)\|_C$$

i.e. in (2.5) the supremum is attained by the function $f_r(\omega, X_v; x)$. The proof of the part of the theorem relative to the $C$ metrics is completed.

Now, let us prove (2.5.16). By the definition

$$\Delta_n^v(W^r H^\omega)_{L_p} = \inf_{X \in (X)_v} \Delta(W^r H^\omega, X)_{L_p}, \quad p > 0. \tag{2.5.17}$$

For each mesh $X \in (X)_v$ we have

$$[\Delta(W^r H^\omega, X)_{L_p}]^p \overset{(2.5.11)}{=} \sum_{i=1}^n \int_{b_{i-1}}^{b_i} \varphi^p(|x - x_i|)dx \overset{(2.1.2)}{=} L_n(\varphi^p; B, X) \tag{2.5.18}$$

where the function $\varphi(u)$ $(u \in [0, 1])$ is determined by (2.5.10).

Taking into account that the function $\varphi^p(u)$, $p > 0$ is increasing, and the mesh pair $(B, X)$ for $X \in (X)_v$ belongs to the set $(A, X)_v$, then for each mesh $x \in (X)_v$ we obtain by the optimization theorem

$$L_n(\varphi^p; B, X) \ge L_n(\varphi^p; A_v, X_v) \overset{(2.5.18)}{=} [\Delta(W^r H^\omega, X_v)_{L_p}]^p \tag{2.5.19}$$

since the mesh $A_v$ is determined by the mesh $X_v$ according to formulas (1.5.21). From (2.5.18) and (2.5.19) it follows that for each mesh $X \in (X)_v$

$$\Delta(W^r H^\omega, X)_{L_p} \ge \Delta(W^r H^\omega, X_v)_{L_p}, \quad p > 0$$

holds. Taking into account that $X \in (X)_v$, from the last inequality we get

$$\inf\{\Delta(W^r H^\omega, X)_{L_p} : X \in (X)_v\} = \Delta(W^r H^\omega, X_v)_{L_p}, \quad p > 0. \quad (2.5.20)$$

From (2.5.17), (2.5.20), (2.5.18), Corollary 2.5.1 and (2.1.3) it follows that

$$\Delta_n^v(W^r H^\omega)_{L_p} = \inf_{X \in (X)_v} \Delta(W^r H^\omega, X)_{L_p}$$

$$= \Delta(W^r H^\omega, X_v)_{L_p} = [L_n(\varphi^p; A_v, X_v)]^{1/p}$$

$$= \sup_{f \in W^r H^\omega} \|f(.) - \Phi^r(f, X_v; .)\|_{L_p}$$

$$= \frac{D^r(A_v, X_v)_\infty}{(r-1)!} \cdot \left\{ \int_0^1 \left[ \int_0^1 u^r (1-z)^{r-1} \cdot \omega(zuD(A_v, X_v)_\infty) dz \right]^p du \right\}^{1/p}.$$

Thus, (2.5.16) is proved. From (2.5.16) follows the optimality of the mesh $X_v$ and the method $\Phi^r(f, X_v; x)$ with respect to all methods of the form (1.5.15), built on meshes $X \in (X)$, for recovery of functions of the classes $W^r H^\omega$ $(r = 2, 4, \ldots)$ and in integral $(L_p, p > 0)$ metrics.

Finally, from (2.5.16) and (2.5.13) we get

$$\sup_{f \in W^r H^\omega} \|f(.) - \Phi^r(f, X_v; .)\|_{L_p} = [L_n(\varphi^p; A_v, X_v)]^{1/p} = \|f_r(.) - \Phi^r(f_r, X_v; .)\|_{L_p}$$

and thus the theorem is completely proved.

We can write explicitly various cases of optimal methods of form (1.5.25) for recovery of functions of the classes $W^r H^\omega$ $(r = 0, 2, 4, 6, \ldots)$ as follows:

$$\Phi^r(f, X_0; x) = \sum_{i=1}^{n} \sum_{k=1}^{r} \chi_i(A_0, x) \frac{f^{(k)}((2i-1)/2n))}{k!} \left( x - \frac{2i-1}{2n} \right)^k \quad (2.5.21_0)$$

$$\Phi^r(f, X_1; x) = \sum_{i=1}^{n} \sum_{k=1}^{r} \chi_i(A_1, x) \frac{f^{(k)}((2i-2)/(2n-1))}{k!} \left( x - \frac{2i-2}{2n-1} \right)^k \quad (2.5.21_1)$$

$$\Phi^r(f, X_2; x) = \sum_{i=1}^{n} \sum_{k=1}^{r} \chi_i(A_2, x) \frac{f^{(k)}((i-1)/(n-1))}{k!} \left( x - \frac{i-1}{n-1} \right)^k \quad (2.5.21_2)$$

$$\Phi^r(f, X_3; x) = \sum_{i=1}^{n} \sum_{k=1}^{r} \chi_i(A_3, x) \frac{f^{(k)}((2i-1)/(2n-1))}{k!} \left( x - \frac{2i-1}{2n-1} \right)^k \quad (2.5.21_3)$$

where $\chi_i(A_v, x)$ is the characteristic function of the interval $[a_{i-1}^{(v)}, a_i^{(v)})$ ($i = 1, 2, \ldots, n$; $v = 0, 1, 2, 3$; $[a_{n-1}^{(v)}, a_n^{(v)}) = [a_{n-1}^{(v)}, a_n^{(v)}]$), determined by the nodes of the mesh $A_v = (a_0^{(v)}, a_1^{(v)}, \ldots, a_n^{(v)})$, defined in $(1.6.5_v)$.

Also, we can give explicitly the errors of these methods in $C$ and $L_p$, ($p > 0$) metrics for $r = 2, 4, 6, \ldots$ (for $r = 0$ they are given in $(2.4.14_v)$ and $(2.4.15_v)$, $v = 0, 1, 2, 3$) as follows:

I   In $C$ metrics:

$$\Delta_n^0(W^r H^\omega)_C = 1/((2n)^r(r-1)!) \int_0^1 (1-z)^{r-1} \cdot \omega(z/2n) \mathrm{d}z; \quad (2.5.22_0)$$

$$\Delta_n^1(W^r H^\omega)_C = 1/((2n-1)^r(r-1)!) \int_0^1 (1-z)^{r-1} \cdot \omega(z/(2n-1)) \mathrm{d}z; \quad (2.5.22_1)$$

$$\Delta_n^2(W^r H^\omega)_C = 1/((2n-2)^r(r-1)!) \int_0^1 (1-z)^{r-1} \cdot \omega(z/(2n-2)) \mathrm{d}z; \quad (2.5.22_2)$$

$$\Delta_n^3(W^r H^\omega)_C = 1/((2n-1)^r(r-1)!) \int_0^1 (1-z)^{r-1} \cdot \omega(z/(2n-1)) \mathrm{d}z. \quad (2.5.22_3)$$

II   In integral metrics:

$$\Delta_n^0(W^r H^\omega)_{L_p} = 1/((2n)^r(r-1)!) \left\{ \int_0^1 \left[ \int_0^1 u^r(1-z)^{r-1}\omega(zu/2n)\mathrm{d}z \right]^p \mathrm{d}u \right\}^{1/p};$$

$$(2.5.23_0)$$

$$\Delta_n^1(W^r H^\omega)_{L_p} = 1/((2n-1)^r(r-1)!) \left\{ \int_0^1 \left[ \int_0^1 u^r(1-z)^{r-1}\omega(zu/(2n-1))\mathrm{d}z \right]^p \mathrm{d}u \right\}^{1/p};$$

$$(2.5.23_1)$$

$$\Delta_n^0(W^r H^\omega)_{L_p} = 1/((2n-2)^r(r-1)!)\left\{\int_0^1\left[\int_0^1 u^r(1-z)^{r-1}\omega(zu/(2n-2))\mathrm{d}z\right]^p \mathrm{d}u\right\}^{1/p};$$

$$(2.5.23_2)$$

$$\Delta_n^3(W^r H^\omega)_{L_p} = 1/((2n-1)^r(r-1)!)\left\{\int_0^1\left[\int_0^1 u^r(1-z)^{r-1}\omega(zu/(2n-1))\mathrm{d}z\right]^p \mathrm{d}u\right\}^{1/p}.$$

$$(2.5.23_3)$$

# 3

---

# Recovery of functions of several variables

In the present chapter the problem for recovery of functions of two variables in the classes $H^{\omega}(D_2)$ and $H^{\omega_1 \omega_2}(D_2)$ in the uniform metric is considered.

## 3.1 Best method for recovery of functions of several variables

Let $n \in \mathbb{N}$, $m \in \mathbb{N}$; let $\omega_1(t)$, $\omega_2(t)$ and $\omega(t)$ be arbitrary moduli of continuity, and $X = (x_1, x_2, \ldots, x_n)$, $X \in (X)_0$, $Y = (y_1, y_2, \ldots, y_m)$, $Y \in (Y)_0$ be arbitrary meshes of the segment $[0, 1]$.

We consider the lattice $(X, Y)$ of the points $M_{ij} = (x_i, y_j)$, $i = 1, 2, \ldots, n$; $j = 1, 2, \ldots, m$ in the unit square $D_2 = \{(x, y): 0 \le x \le 1, 0 \le y \le 1\}$. We say that the lattice $(X, Y)$ belongs to the set $(X, Y)_{v,\mu}$ and we write $(X, Y) \in (X, Y)_{v,\mu}$ provided $X \in (X)_v$ and $Y \in (Y)_\mu$.

We suppose that for each function $f(x, y)$ in the points of the lattice $(X, Y)$ are known, i.e. we suppose that the information

$$T(F; X, Y) = (f(x_i, y_j))_{i=1, j=1}^{n, m} \tag{3.1.1}$$

is known.

For each fixed point $(x, y) \in D_2$ we consider the problem for recovery of the functional $Lf = f(x, y)$ in the class $H^{\omega}(D_2)$ or in the class $H^{\omega_1 \omega_2}(D_2)$ on the information (3.1.1) in the uniform metric.

Let $S_{12}$ (or $S_0$) be the set of all possible methods for recovery of the functional $Lf = f(x, y)$ in the class $H^{\omega_1 \omega_2}(D_2)$ (or $H^{\omega}(D_2)$) on the information (3.1.1).

### Lemma 3.1.1

*The classes $H^{\omega_1 \omega_2}(D_2)$ and $H^\omega(D_2)$ for fixed moduli of continuity $\omega_1(t)$, $\omega_2(t)$ and $\omega(t)$ are convex sets in the normed linear space $C(D_2)$.*

### Proof

Let $f$ and $\varphi$ be functions of the class $H^{\omega_1 \omega_2}(D_2)$ and $\lambda$ be an arbitrary number of the segment $[0, 1]$. Then, for every two points $(x, y)$ and $(x', y')$ of $D_2$, due to (1.4.9) the inequalities

$$\left| f(x, y) - f(x', y') \right| \leq \omega_1(|x - x'|) + \omega_2(|y - y'|)$$
$$\left| \varphi(x, y) - \varphi(x', y') \right| \leq \omega_1(|x - x'|) + \omega_2(|y - y'|) \tag{3.1.2}$$

hold.

Now, for the function

$$g(x, y) = \lambda f(x, y) + (1 - \lambda)\varphi(x, y) \tag{3.1.3}$$

and for every two points $(x, y)$ and $(x', y')$ of $D_2$ we obtain

$$|g(x, y) - g(x', y')| \leq \lambda |f(x, y) - f(x', y')| + (1 - \lambda)|\varphi(x, y) - \varphi(x', y')|$$
$$\leq \lambda[\omega_1(|x - x'|) + \omega_2(|y - y'|)]$$
$$+ (1 - \lambda)[\omega_1(|x - x'|) + \omega_2(|y - y'|)]$$
$$= \omega_1(|x - x'|) + \omega_2(|y - y'|).$$

This relation shows that the function (3.1.3) belongs to the set $H^{\omega_1 \omega_2}(D_2)$, i.e. that it is a convex subset of $C(D_2)$.

Similarly, from the inequalities

$$|f(M) - f(M')| \leq \omega(\rho(M, M'))$$

and

$$|\varphi(M) - \varphi(M')| \leq \omega(\rho(M, M'))$$

we obtain

$$|g(M) - g(M')| \leq \lambda |f(M) - f(M')| + (1 - \lambda) \cdot |\varphi(M) - \varphi(M')|$$

$$\leq \lambda \omega(\rho(M, M')) + (1 - \lambda)\omega(\rho(M, M') = \omega(\rho(M, M')).$$

The last relation shows that $f \in H^{\omega}(D_2)$ and $\varphi \in H^{\omega}(D_2)$ imply $\lambda f + (1 - \lambda)$ $\varphi \in H^{\omega}(D_2)$, $0 \leq \lambda \leq 1$, i.e. the convexity of the class $H^{\omega}(D_2)$ in $C(D_2)$.

The lemma is proved.


Since the first of the inequalities (3.1.2) implies

$$|(-1)f(x, y) - (-1)f(x', y')| \leq \omega_1(|x - x'|) + \omega_2(|y - y'|)$$

then $f \in H^{\omega_1 \omega_2}(D_2)$ implies $(-1) \cdot f \in H^{\omega_1 \omega_2}(D_2)$, i.e. the set $H^{\omega_1 \omega_2}(D_2)$ is centrally symmetric in $C(D_2)$. In the same way one can show that the set $H^{\omega}(D_2)$ is centrally symmetric in $C(D_2)$.

Further, we use the sets

$$H^{\omega_1 \omega_2}(D_2)_0 = \{f(x, y) : f \in H^{\omega_1 \omega_2}(D_2), f(x_i, y_j) = 0,$$

$$i = 1, 2, \ldots, n; j = 1, 2, \ldots, m\} = \{f(x, y) : f \in H^{\omega_1 \omega_2}(D_2), T(f; X, Y) = 0\}$$

$$(3.1.4)$$

and

$$H^{\omega}(D_2)_0 = \{f(x, y) : f \in H^{\omega}(D_2), T(f; X, Y) = 0\}. \qquad (3.1.5)$$


Using formulas (1.5.21), the mesh $B = (b_0, b_1, \ldots, b_n)$ is uniquely determined by the mesh $X = (x_1, x_2, \ldots, x_n)$. The formulas

$$e_0 = 0, \qquad e_m = 1, \qquad e_j = \frac{y_j + y_{j+1}}{2} \quad (j = 1, 2, \ldots, m - 1) \quad (3.1.6)$$

determine uniquely a mesh $E = (e_0, e_1, e_2, \ldots, e_m)$ on the mesh $Y = (y_1, y_2, \ldots, y_m)$.

Let $\chi_i(x)$ and $\chi_j(y)$ be the characteristic functions of the intervals $[b_{i-1}, b_i)$ and $[e_{j-1}, e_j)$ $(i = 1, 2, \ldots, n; j = 1, 2, \ldots, m)$ respectively. (Let us recall the convention: $[\alpha, \beta) = [\alpha, \beta]$ if $\beta = 1$, and $\alpha \in [0, 1]$.)

The function

$$f_{12}(X, Y; x, y) = \sum_{i=1}^{n} \sum_{j=1}^{m} \chi_i(x)\chi_j(y)[\omega_1(|x - x_i|) + \omega_2(|y - y_j|)]$$

$$= \sum_{i=1}^{n} \chi_i(x)\omega_1(|x - x_i|) + \sum_{j=1}^{n} \chi_j(y)\omega_2(|y - y_j|) \qquad (3.1.7)$$

of $(x, y) \in D_2$ is said to be the standard function for the class $H^{\omega_1 \omega_2}(D_2)$ on the lattice, and the function

$$f_0(X, Y, x, y) = \sum_{i=1}^{n} \sum_{j=1}^{m} \chi_i(x)\chi_j(y)\omega(\rho(M, M_{ij})) \tag{3.1.8}$$

of $(x, y) \in D_2$, where $\rho(M, M_{ij})$ is the Euclidean distance between the points $M = (x, y) \in D_2$ and $M_{ij} = (x_i, y_j) \in (X, Y)$ is said to be the standard function for the class $H^\omega(D_2)$ and the lattice $(X, Y)$.

Similarly to Lemma 2.2.1, we can prove the following lemma.

### Lemma 3.1.2

*Let $n \in \mathbb{N}$, $m \in \mathbb{N}$; let $(X, Y) \in (X, Y)_{0,0}$ be an arbitrary lattice of points of the unit square $D_2$ and $\omega_1(t)$, $\omega_2(t)$ and $\omega(t)$ be arbitrary moduli of continuity. Then the standard functions (3.1.7) and (3.1.8) have the following properties:*

*(i)*   $f_{12}(X, Y; x, y) \geq 0$, $(x, y) \in D_2$;
*(ii)*   $f_{12}(X, Y; x_i, y_j) = 0$ $(i = 1, 2, \ldots, n; j = 1, 2, \ldots, m)$;
*(iii)*   $f_{12}(X, Y; x, y) \in C(D_2)$;
*(iv)*   $f_{12}(X, Y; x, y) \in H^{\omega_1 \omega_2}(D_2)$;
*(v)*   $f_{12}(X, Y; x, y) \in H^{\omega_1 \omega_2}(D_2)_0$;
*(vi)*   *for every function $f \in H^{\omega_1 \omega_2}(D_2)_0$ and for each point $(x, y) \in D_2$ the inequality*

$$f(x, y) \leq f_{12}(X, Y; x, y) \tag{3.1.9}$$

*holds. This means that for each point $(x, y) \in D_2$ the equality*

$$\sup\{f(x, y): f \in H^{\omega_1 \omega_2}(D_2)_0\} = f_{12}(X, Y; x, y) \tag{3.1.10}$$

*is satisfied.*

*Similar relations hold for the function $f_0(X, Y; x, y)$ also:*

*(i′)*   $f_0(X, Y; x, y) \geq 0$,   $(x, y) \in D_2$;
*(ii′)*   $f_0(X, Y; x_i, y_j) = 0$   $(i = 1, 2, \ldots, n; j = 1, 2, \ldots, m)$;
*(iii′)*   $f_0(X, Y; x, y) \in C(D_2)$;
*(iv′)*   $f_0(X, Y; x, y) \in H^\omega(D_2)$;
*(v′)*   $f_0(X, Y; x, y) \in H^\omega(D_2)_0$;
*(vi′)*   *for every function $f \in H^\omega(D_2)_0$ and for each point $(x, y) \in D_2$ the inequality*

$$f(x, y) \leq f_0(X, Y; x, y) \tag{3.1.11}$$

*holds. This means that for each point $(x, y) \in D_2$ the equation*

$$\sup\{f(x, y): f \in H^\omega(D_2)_0\} = f_0(X, Y; x, y) \tag{3.1.12}$$

*holds.*

### Lemma 3.1.3

*Let $n \in \mathbb{N}$, $m \in \mathbb{N}$, $(X, Y) \in (X, Y)_{0,0}$ be an arbitrary point lattice in th unit square $D_2$, and $\omega_1(t)$, $\omega_2(t)$ and $\omega(t)$ be arbitrary moduli of continuity.*

*Then, the errors of the best method for recovery of the functional $Lf = f(x, y)$ in the classes $H^{\omega_1\omega_2}(D_2)$ and $H^\omega(D_2)$ on the information $T(f; X, Y)$ are given by the corresponding formulas*

$$R(H^{\omega_1\omega_2}(D_2); X, Y; x, y) = f_{12}(X, Y; x, y), \quad (x, y) \in D_2, \qquad (3.1.13)$$

*and*

$$R(H^\omega(D_2); X, Y; x, y) = f_0(X, Y; x, y), \quad (x, y \in D_2 \qquad (3.1.14)$$

*where $f_{12}(X, Y; x, y)$ and $f_0(X, Y; x, y)$ are the standard functions (3.1.7) and (3.1.8), respectively.*

### Proof

As proved above, the sets $H^{\omega_1\omega_2}(D_2)$ and $H^\omega(D_2)$ are centrally symmetric in the normed linear space $C(D_2)$. From the hypothesis of Smolyak's lemma it remains to verify the inequalities:

$$\sup\{Lf : f \in H^{\omega_1\omega_2}(D_2)_0\} < +\infty, \quad (x, y) \in D_2$$

and

$$\sup\{Lf : f \in H^\omega(D_2)_0\} < +\infty, \quad (x, y) \in D_2.$$

Let $(x, y) \in D_2$ be an arbitrary fixed point. By the properties (v) and (vi) of the function $f_{12}(X, Y; x, y)$ of Lemma 3.1.2, we have

$$\sup\{Lf : f \in H^{\omega_1\omega_2}(D_2)_0\}$$

$$= \sup\{f(x, y) : f \in H^{\omega_1\omega_2}(D_2)_0\}$$

$$\overset{(3.1.10)}{=} f_{12}(X, Y; x, y) \overset{(3.1.7)}{=} \sum_{i=1}^n \chi_i(x)\omega_1(|x - x_i|)$$

$$+ \sum_{j=1}^n \chi_j(y)\omega_2(|y - y_j|) \leqq \omega_1(1) + \omega_2(1) < +\infty$$

and thus the first of the above two inequalities is proved.

The second inequality can be proved in the same way. Using properties (v′) and (vi′) of the function $f(X, Y; x, y)$ of Lemma 3.1.2, we have

$$\sup\{Lf : f \in H^{\omega}(D_2)_0\}$$

$$= \sup\{f(x, y) : f \in H^{\omega}(D_2)_0\} \overset{(3.1.12)}{=} f_0(X, Y; x, y)$$

$$\overset{(3.1.8)}{=} \sum_{i=1}^{n} \sum_{j=1}^{m} \chi_i(x)\chi_j(y)\omega(\rho(M, M_{ij}))$$

$$\leqq \omega(\sqrt{2}) < +\infty$$

and thus the second of the above two inequalities is proved.

According to the corollary of Smolyak's lemma, we have

$$R(H^{\omega_1\omega_2}(D_2); X, Y; x, y) \overset{\text{def}}{=} \inf_{\Phi \in S_{12}} \sup_{f \in H^{\omega_1\omega_2}(D_2)} |Lf - \Phi|$$

$$= \sup\{Lf : f \in H^{\omega_1\omega_2}(D_2)_0\}$$

$$= \sup\{f(x, y) : f \in H^{\omega_1\omega_2}(D_2)_0\} \overset{(3.1.10)}{=} f_{12}(X, Y; x, y), \quad (x, y) \in D_2$$

and relation (3.1.13) is proved.

Analogously, again from the corollary of Smolyak's lemma, we have

$$R(H^{\omega}(D_2); X, Y; x, y) = \inf_{\Phi \in S_0} \sup_{f \in H^{\omega}(D_2)} |Lf - \Phi| = \sup\{Lf : f \in H^{\omega}(D_2)_0\}$$

$$= \sup\{f(x, y) : f \in H^{\omega}(D_2)_0\} = f_0(X, Y; x, y).$$

Thus relation (3.1.14) is proved.

### Theorem 3.1.1

*Let n and m be positive integers, $(X, Y) \in (X, Y)_{0,0}$ be an arbitrary lattice of nodes $M_{ij} = (x_i, y_j)$ $(i = 1, 2, \ldots, n; j = 1, 2, \ldots, m)$ in the unit square $D_2$ and $\omega_1(t)$, $\omega_2(t)$ and $\omega(t)$ be arbitrary moduli of continuity. Then, out of all possible methods for recovery of the functional $Lf = f(x, y)$ in the classes $H^{\omega_1\omega_2}(D_2)$ and $H^{\omega}(D_2)$ on the information $T(f; X, Y)$ the best is the method*

$$\Phi(f; X, Y; x, y) = \sum_{i=1}^{n} \sum_{j=1}^{m} \chi_i(x)\chi_j(y)f(x_i, y_j) \qquad (x, y) \in D_2 \qquad (3.1.15)$$

*where $\chi_i(x)$ and $\chi_j(y)$ are the corresponding characteristic functions of the intervals $[b_{i-1}, b_i]$ $(i = 1, 2, \ldots, n)$ and $[e_{j-1}, e_j]$, $(j = 1, 2, 3, \ldots, m)$ determined*

*by the meshes $B = (b_0, b_1, b_2, \ldots, b_n)$ and $E = (e_0, e_1, e_2, \ldots, b_m)$ introduced in
(1.5.21) and (3.1.6).*

*Then for each point $(x, y) \in D_2$ the relations*

$$R(H^{\omega_1 \omega_2}(D_2); X, Y; x, y)$$

$$= f_{12}(X, Y; x, y)$$

$$= \sup\{|f(x, y) - \Phi(f; X, Y; x, y)| : f \in H^{\omega_1 \omega_2}(D_2)\} \qquad (3.1.16)$$

*and*

$$R(H^{\omega}(D_2); X, Y; x, y)$$

$$= f_0(X, Y; x, y)$$

$$= \sup\{|f(x, y) - \Phi(f; X, Y; x, y)| : f \in H^{\omega}(D_2)\} \qquad (3.1.17)$$

*hold, where*

$$|f_{12}(X, Y; x, y) - \Phi(f_{12}; X, Y; x, y)| = f_{12}(X, Y; x, y) \qquad (3.1.18)$$

*and*

$$|f_0(X, Y; x, y) - \Phi(f_0; X, Y; x, y)| = f_0(X, Y; x, y) \qquad (3.1.19)$$

*where $f_{12}$ and $f_0$ are the standard functions (3.1.7) and (3.1.8).*

### Proof

For every fixed point $(x, y) \in D_2$ and for each function $f \in H^{\omega_1 \omega_2}(D_2)$ we have

$$|f(x, y) - \Phi(f; X, Y; x, y)|$$

$$\overset{(3.1.15)}{=} \left| \sum_{i=1}^{n} \sum_{j=1}^{m} \chi_i(x)\chi_j(y)[f(x, y) - f(x_i, y_j)] \right|$$

$$\leq \sum_{i=1}^{n} \sum_{j=1}^{m} \chi_i(x)\chi_j(y)[\omega_1(|x - x_i|) + \omega_2(|y - y_j|)]$$

$$\overset{(3.1.7)}{=} f_{12}(X, Y; x, y).$$

Therefore

$$\sup\{|f(x, y) - \Phi(f; X, Y; x, y)| : f \in H^{\omega_1 \omega_2}(D_2)\} \leq f_{12}(X, Y; x, y). \qquad (3.1.20)$$

Keeping in mind (3.1.13) and the fact that method (3.1.15) belongs to $S_{12}$, we obtain the inequality

$$f_{12}(X, Y; x, y) = R(H^{\omega_1\omega_2}(D_2); X, Y; x, y)$$

$$\leq \sup_{f \in H^{\omega_1\omega_2}(D_2)} |f(x, y) - \Phi(f; X, Y; x, y)|.$$

From this inequality and from (3.1.20) the relation (3.1.16) follows. This, according to Definition (1.5.3) means that method (3.1.15) is the best in the class $H^{\omega_1\omega_2}(D_2)$ on the information $T(f; X, Y)$.

From properties (i), (ii) and (iv) of the function $f_{12}(X, Y; x, y)$ (see Lemma 3.1.2) and from (3.1.15) the equality (3.1.18) follows.

Further, for each fixed point $(x, y) \in D_2$ and for every function $f(x, y) \in H^{\omega}(D_2)$ we obtain the estimation

$$|f(x, y) - \Phi(f; X, Y; x, y)| \overset{(3.1.15)}{=} \left| \sum_{i=1}^{n} \sum_{j=1}^{m} \chi_i(x)\chi_j(y)[f(x, y) - f(x_i, y_j)] \right|$$

$$\leq \sum_{i=1}^{n} \sum_{j=1}^{m} \chi_i(x)\chi_j(y)\omega(\rho(M, M_{ij}))$$

$$\overset{(3.1.8)}{=} f_0(X, Y; x, y).$$

Therefore

$$\sup_{f \in H^{\omega}(D_2)} |f(x, y) - \Phi(f; X, Y; x, y)| \leq f_0(X, Y; x, y). \tag{3.1.21}$$

Taking into account (3.1.14) and the fact that method (3.1.15) belongs to $S_0$, we obtain the inequality

$$f_0(X, Y; x, y) = R(H^{\omega}(D_2); X, Y; x, y) \leq \sup_{f \in H^{\omega}(D_2)} |f(x, y) - \Phi(f; X, Y; x, y)|.$$

From this inequality and from (3.1.21) relation (3.1.17) follows. According to Definition (1.5.3), this means that method (3.1.15) is the best in the class $H^{\omega}(D_2)$ on the information $T(f; X, Y)$.

From properties (i'), (ii') and (iv') of the function $f_0(X, Y; x, y)$ (see Lemma 3.1.2) and from (3.1.15) the equality (3.1.18) follows. The theorem is proved.

For the errors (see (1.5.12))

$$R(H^{\omega_1\omega_2}(D_2); X, Y)_{C(D_2)} \overset{\text{def}}{=} \|R(H^{\omega_1\omega_2}(D_2); X, Y; ., ., .)\|_{C(D_2)} \tag{3.1.22}$$

and

$$R(H^\omega(D_2); X, Y)_{C(D_2)} \overset{\text{def}}{=} \|R(H^\omega(D_2); X, Y; .,.)\|_{C(D_2)} \qquad (3.1.23)$$

of the best method in uniform metrics in $H^{\omega_1\omega_2}(D_2)$ and $H^\omega(D_2)$ respectively, according to (3.1.16) and (3.1.17) we have

$$R(H^{\omega_1\omega_2}(D_2); X, Y)_{C(D_2)} = \sup_{(x,y)\in D_2} \sup_{f\in H^{\omega_1\omega_2}(D_2)} |f(x,y) - \varphi(f; X, Y; x, y)|$$

$$= \sup_{(x,y)\in D_2} f_{12}(X, Y; x, y)$$

$$= \sup_{f\in H^{\omega_1\omega_2}(D_2)} \|f(.,.) - \Phi(f; X, Y; .,.)\|_{C(D_2)} \qquad (3.1.24)$$

and

$$R(H^\omega(D_2); X, Y)_{C(D_2)} = \sup_{(x,y)\in D_2} \sup_{f\in H^\omega(D_1)} |f(x,y) - \Phi(f; X, Y; x, y)]$$

$$= \sup_{(x,y)\in D_2} f_0(X, Y; x, y)$$

$$= \sup_{f\in H^\omega(D_2)} \|f(.,.) - \Phi(f; X, Y; .,.)\|_{C(D_2)}. \qquad (3.1.25)$$

In the last equalities of (3.1.24) and (3.1.25) we used the possibility to interchange the order of the computing of the supremums (see [6], p 304, D17).

Let us now calculate $\sup_{D_2} f_{12}(X, Y; x, y)$ and $\sup_{D_2} f_0(X, Y; x, y)$.

Let $(x, y)$ be an arbitrary point of the unit square $D_2$. Then it belongs to some of the rectangles† $D_{ij} = \{(x, y): x\in[b_{i-1}, b_i)\ y\in[e_{j-1}, e_j)\ (i=1, 2, \ldots, n;\ j=1, 2, \ldots, m)$. For example, let $(x, y)\in D_{p,q}$, $1\le p\le n$, $1\le q\le m$. Then

$$f_{12}(X, Y; x, y) = \omega_1(|x - x_p|) + \omega_2(|y - y_q|). \qquad (3.1.26)$$

From the monotony of the moduli of continuity $\omega_1(t)$, $\omega_2(t)$ and from (see (1.6.14))

$$|x - x_p| \le D(B, X)_\infty, \quad |y - y_q| \le D(E, Y)_\infty$$

where $D(A, X)_\infty$ is the uniform deviation of the mesh pair $(A, X)$, the inequalities

$$\omega_1(|x - x_p|) \le \omega_1(D(B, X)_\infty), \qquad \omega_2(|y - y_q|) \le \omega_2(D(E, Y)_\infty)$$

† If $\beta = 1$, then by definition $[\alpha, \beta) = [\alpha, \beta]$ when $\beta = 1$ and $\alpha\in[0, 1]$.

follow. From them and (3.1.26) the estimate

$$\|f_{12}(X, Y; ., .)\|_{C(D_2)} \leqq \omega_1(D(B, X)_\infty) + \omega_2(D(E, Y)_\infty) \qquad (3.1.27)$$

follows.

For example, let

$$D(B, X)_\infty = |b_{s-1} - x_s|, \quad D(E, Y)_\infty = |e_{\tau-1} - y_\tau|.$$

Then, from the estimate obtained and from (3.1.26) the relation

$$f_{12}(X, Y; b_{s-1}, e_{\tau-1}) = \omega_1(D(B, X)_\infty) + \omega_2(D(E, Y)_\infty)$$

follows. This, along with (3.1.27) gives

$$\|f_{12}(X, Y; ., .)\|_{C(D_2)} = \omega_1(D(B, X)_\infty) + \omega_2(D(E, Y)_\infty). \qquad (3.1.28)$$

In a similar way one can find

$$\|f_0(X, Y; ., .)\|_{C(D_2)} = \omega(\sqrt{D^2(B, X)_\infty + D^2(E, Y)_\infty}). \qquad (3.1.29)$$

Thus we have proved the following assertion:

### Theorem 3.1.2

*Let $n$ and $m$ be positive integers, $(X, Y) \in (X, Y)_{0,0}$ be an arbitrary lattice of nodes $M_{ij} = (x_i, y_j)$, $(i = 1, 2, \ldots, n; j = 1, 2, \ldots, m)$ in the unit square $D_2$, and $\omega_1(t)$, $\omega_2(t)$ and $\omega(t)$ be arbitrary moduli of continuity.*

*From all possible methods for recovery of functions of the classes $H^{\omega_1\omega_2}(D_2)$ and $H^\omega(D_2)$ on the information $T(f; X, Y)$ the best one in the uniform metrics is (3.1.15) with the corresponding error*

$$R(H^{\omega_1\omega_2}(D_2); X, Y)_{C(D_2)} = \sup_{f \in H^{\omega_1\omega_2}(D_2)} |f(., .) - \Phi(f; (X, Y; ., .)\|_{C(D_2)}$$

$$= \|f_{12}(X, Y; ., .) - \Phi(f_{12}; X, Y; ., .)\|_{C(D_2)}$$

$$= \|f_{12}(X, Y; ., .)\|_{C(D_2)}$$

$$= \omega_1(D(B, X)_\infty) + \omega_2(D(E, Y)_\infty) \qquad (3.1.30)$$

*and*

$$R(H^\omega(D_2); X, Y)_{C(D_2)} = \sup_{f \in H^\omega(D_2)} \|f(., .) - \Phi(f; X, Y; ., .)\|_{C(D_2)}$$

$$= \|f_0(X, Y; ., .)\|_{C(D_2)}$$

$$= \| f_0(X, Y; ., .) - \Phi(f_0; X, Y; ., .) \|_{C(D_2)}$$

$$= \omega(\sqrt{D^2(B, X)_\infty + D^2(E, Y)_\infty}). \qquad (3.1.31)$$

---

**Remark.** Equations (3.1.30) and (3.1.31) show that the supremums in them are reached by the standard functions (3.1.7) and (3.1.8), respectively.

---

### Proof

Equation (3.1.30) follows from (3.1.24), (3.1.28) and (3.1.18); (3.1.31) follows from (3.1.25), (3.1.29) and (3.1.19).

## 3.2  An optimal method for recovery of functions of several variables

This section is devoted to the optimal methods of recovery of functions of the classes $H^{\omega_1\omega_2}(D_2)$ and $H^\omega(D_2)$ on information $T(f; X, Y)$ in uniform metrics. This is done both in free and restricted lattice of nodes $(X, Y)$ subjected to some additional restrictions.

### Lemma 3.2.1

*Let $P$ and $Q$ be arbitrary sets, and the functions $f: P \to \mathbb{R}$ and $g: Q \to \mathbb{R}$ be bounded from below. Then*

$$\inf_{(x,y)\in P \times Q} [f(x) + g(y)] = \inf_{x\in P} f(x) + \inf_{y\in Q} g(y). \qquad (3.2.1)$$

### Proof

Since the functions $f: P \to \mathbb{R}$ and $g: Q \to \mathbb{R}$ are bounded from below, then there exist the infimums

$$\inf\{f(x): x\in P\} = \alpha \qquad (3.2.2)$$

and

$$\inf\{g(y): y\in Q\} = \beta. \qquad (3.2.3)$$

Let $\varepsilon > 0$ be arbitrarily chosen. From (3.2.2) and (3.2.3) it follows the existence of an element $x_\varepsilon \in P$, such that

$$f(x_\varepsilon) < \alpha + (\varepsilon/2)$$

and an element $y_\varepsilon \in Q$, such that

$$g(y_\varepsilon) < \beta + (\varepsilon/2).$$

Hence there is an element $(x_\varepsilon, y_\varepsilon) \in P \times Q$
$$f(x_\varepsilon) + g(y_\varepsilon) < \alpha + \beta + \varepsilon. \tag{3.2.4}$$

Let $(x, y)$ be an arbitrary element of $P \times Q$. Then from (3.2.2) and (3.2.3)

$$f(x) \geq \alpha \qquad \text{and} \quad g(y) \geq \beta.$$

Hence, for an arbitrary element $(x, y) \in P \times Q$ we have

$$f(x) + g(y) \geq \alpha + \beta. \tag{3.2.5}$$

From (3.2.4) and (3.2.5) the equality

$$\inf_{(x,y) \in P \times Q} [f(x) + g(y)] = \alpha + \beta$$

follows. Along with (3.2.2) and (3.2.3), it gives (3.2.1).

### Theorem 3.2.1

*Let $n$ and $m$ be fixed positive integers, $(v, \mu)$ be an arbitrary element of the set $\{0, 1, 2, 3\} \times \{0, 1, 2, 3\}$, $(X, Y)$ be a lattice of nodes $M_{ij} = (x_i, y_j)$ $(i = 1, 2, \ldots, n; j = 0, 1, 2, \ldots, m)$ in the unit square $D_2$, and $\omega_1(t)$, $\omega_2(t)$ and $\omega(t)$ be arbitrary moduli of continuity. Then out of all methods for recovery of functions of the classes $H^{\omega_1 \omega_2}(D_2)$ and $H^\omega(D_2)$ on information $T(f; X, Y)$, $(X, Y) \in (X, Y)_{v, \mu}$ in uniform metrics, the optimal is the method*

$$\Phi(f; X_v, Y_\mu; x, y) = \sum_{i=1}^{n} \sum_{j=1}^{m} \chi_i^v(x) \chi_j^\mu(y) f(x_i^{(v)}, y_j^{(\mu)}) \tag{3.2.6}$$

*where the mesh $X_v = (x_1^{(v)}, \ldots, x_n^{(v)})$ is determined by (1.6.5$_v$); the mesh $Y_\mu = (y_1^{(\mu)}, \ldots, y_m^{(\mu)}) = X_\mu$ is determined by (1.6.5$_\mu$) for $n = m$; $\chi_s^{(p)}(t)$ is the characteristic function of the interval $[a_{s-1}^{(p)}, a_s^{(p)})$, determined by the mesh*

$A_p = (a_0^{(p)}, a_1^{(p)}, \ldots, a_\tau^{(p)})$ $(\tau = n$ *or* $m, p \in \{0, 1, 2, 3\}$ *defined in* $(1.6.5_p)$. *Then*

$$R_{nm}^{\nu\mu}(H^{\omega_1\omega_2}(D_2))_{C(D_2)} = \inf_{(X,Y)\in(X,Y)_{\nu,\mu}} R(H^{\omega_1\omega_2}(D_2); X, Y)_{C(D_2)}$$

$$= \sup\{\|f(.,.) - \Phi(f; X, Y; ., .)\|_{C(D_2)} : f \in H^{\omega_1\omega_2}(D_2)\}$$

$$= \|f_{12}(X_\nu, Y_\mu; ., .) - \Phi(f_{12}; X_\nu, Y_\mu; ., .)\|_{C(D_2)}$$

$$= \|f_{12}(X_\nu, Y_\mu; ., .)\|_{C(D_2)}$$

$$= \omega_1(D(A_\nu, X_\nu)_\infty) + \omega_2(D(A_\mu, Y_\mu)_\infty) \tag{3.2.7}$$

*and*

$$R_{nm}^{\nu\mu}(H^{\omega}(D_2))_{C(D_2)} = \inf_{(X,Y)\in(X,Y)_{\nu,\mu}} R(H^{\omega}(D_2); X, Y)_{C(D_2)}$$

$$= \sup\{\|f(.,.) - \Phi(f; X_\nu, Y_\mu; ., .)\|_{C(D_2)} : f \in H^{\omega}(D_2)\}$$

$$= \|f_0(X_\nu, Y_\mu; ., .) - \Phi(f_0; X_\nu, Y_\mu; ., .)\|_{C(D_2)}$$

$$= \|f_0(X_\nu, Y_\mu; ., .)\|_{C(D_2)}$$

$$= \omega(\sqrt{D^2(A_\nu, X_\nu)_\infty + D^2(A_\mu, Y_\mu)_\infty}). \tag{3.2.8}$$

### Proof

By definition (1.5.15) (for the two-dimensional case) we get

$$R_{nm}^{\nu\mu}(H^{\omega_1\omega_2}(D_2))_{C(D_2)} = \inf\{R(H^{\omega_1\omega_2}(D_2); X, Y)_{C(D_2)} : (X, Y) \in (X, Y)_{\nu,\mu}\}$$

and according to (3.1.30)

$$\inf_{(X,Y)\in(X,Y)_{\nu,\mu}} R(H^{\omega_1\omega_2}(D_2); X, Y)_{C(D_2)} = \inf_{(X,Y)\in(X,Y)_{\nu,\mu}} [\omega_1(D(B, X)_\infty + \omega_2(D(E, Y)_\infty)]$$

and according to Lemma 3.2.1

$$\inf_{(X,Y)\in(X,Y)_{\nu,\mu}} [\omega_1(D(B, X)_\infty) + \omega_2(D(E, Y)_\infty)]$$

$$= \inf_{X\in(X)_\nu} \omega_1(D(B, X)_\infty) + \inf_{Y\in(Y)_\mu} \omega_2(D(E, Y)_\infty)$$

and according to Lemma 2.4.1

$$\inf_{X\in(X)_\nu} \omega_1(D(B, X)_\infty) = \omega_1(\inf_{X\in(X)_\nu} D(B, X)_\infty)$$

$$\inf_{Y\in(Y)_\mu}\omega_2(D(E,Y)_\infty)=\omega_2(\inf_{Y\in(Y)_\mu}D(E,Y)_\infty)$$

and, finally, according to (2.1.19) we have

$$\inf_{X\in(X)_v}D(B,X)_\infty=D(A_v,X_v)_\infty$$

$$\inf_{Y\in(Y)_\mu}D(E,Y)_\infty=D(A_\mu,Y_\mu)_\infty.$$

(3.2.9)

These equations imply the relation

$$R_{nm}^{v\mu}(H^{\omega_1\omega_2}(D_2))_C=\omega_1(D(A_v,X_v)_\infty)+\omega_2(D(A_\mu,Y_\mu)_\infty).$$

Using Theorem 3.1.2 (equations (3.1.30)) from the last relation we obtain equations (3.2.7). Thus all the assertions of the theorem for the class $H^{\omega_1\omega_2}(D_2)$ are proved.

The part of the theorem concerning the class $H^\omega(D_2)$ can be proved in an analogous manner. Namely, by Definition (1.5.15) we get

$$R_{nm}^{v\mu}(H^\omega(D_2))_{C(D_2)}=\inf_{(X,Y)\in(X,Y)_{v,\mu}}R(H^\omega(D_2);X,Y)_{C(D_2)}$$

and by (3.1.31)

$$\inf_{(X,Y)\in(X,Y)_{v,\mu}}R(H^\omega(D_2);X,Y)_{C(D_2)}=\inf_{(X,Y)\in(X,Y)_{v,\mu}}\omega(\sqrt{D^2(B,X)_\infty+D^2(E,Y)_\infty}).$$

Taking into account that the function $t^{1/2}$ is a modulus of continuity and the almost evident assertion that the composition $\omega(t^{1/2})$ of two moduli of continuity is a modulus of continuity also, we get the relation

$$\inf_{(X,Y)\in(X,Y)_{v,\mu}}\omega(\sqrt{D^2(B,X)_\infty+D^2(E,Y)_\infty})$$

$$=\omega\{\sqrt{\inf_{(X,Y)\in(X,Y)_{v,\mu}}[D^2(B,X)_\infty+D^2(E,Y)_\infty]}\}$$

by interchanging inf and the modulus of continuity.

Now, using Lemma 3.2.1 and equations (3.2.9), from the yet-to-be-obtained relations it follows that

$$R_{nm}^{v\mu}(H^\omega(D_2))_{C(D_2)}=\omega(\sqrt{D^2(A_v,X_v)_\infty+D^2(A_\mu,Y_\mu)_\infty}).$$

Again using theorem 3.1.2 (equations (3.1.31)), from the last equation the equations (3.1.8) follow. Thus all the assertions of the theorem for the class of functions $H^{\omega}(D_2)$ are proved.

Thus Theorem 3.2.1 is completely proved.

In table 3.2.1 we give in an explicit form the assertions of Theorem 3.2.1. Each row of this table contains the assertion of the theorem for the corresponding case $(v,\mu)$, $v\in\{0,1,2,3\}$, $\mu\in\{0,1,2,3\}$. For the optimal method: $v$; $\mu$; the restrictions on $n$ and $m$ (if there are such); $\Phi(f;X_v,Y_\mu;x,y)$; the error and the extremal for the class $H^{\omega_1\omega_2}(D_2)$ along with the extremal for the class are given.

Since for practical applications it is essential that

$$f(x,y) \approx \Phi(f;X_v,Y_\mu;x,y)$$

we give the explicit expressions for $\Phi(f;X_v,Y_\mu;x,y)$ as follows:

$$\Phi(f;X_0,Y_0;x,y) = \sum_{i=1}^{n}\sum_{j=1}^{m} \chi_i(A_0;x)\chi_j(A_0;y)f\left(\frac{2i-1}{2n},\frac{2j-1}{2m}\right);$$

$$\Phi(f;X_1,Y_0;x,y) = \sum_{i=1}^{n}\sum_{j=1}^{m} \chi_i(A_1;x)\chi_j(A_0;y)f\left(\frac{2i-2}{2n-1},\frac{2j-1}{2m}\right);$$

$$\Phi(f;X_2,Y_0;x,y) = \sum_{i=1}^{n}\sum_{j=1}^{m} \chi_i(A_2;x)\chi_j(A_0;y)f\left(\frac{i-1}{n-1},\frac{2j-1}{2m}\right);$$

$$\Phi(f;X_3,Y_0;x,y) = \sum_{i=1}^{n}\sum_{j=1}^{m} \chi_i(A_3;x)\chi_j(A_0;y)f\left(\frac{2i-1}{2n-1},\frac{2j-1}{2m}\right);$$

$$\Phi(f;X_0,Y_1;x,y) = \sum_{i=1}^{n}\sum_{j=1}^{m} \chi_i(A_0;x)\chi_j(A_1;y)f\left(\frac{2i-1}{2n},\frac{2j-2}{2m-1}\right);$$

$$\Phi(f;X_1,Y_1;x,y) = \sum_{i=1}^{n}\sum_{j=1}^{m} \chi_i(A_1;x)\chi_j(A_1;y)f\left(\frac{2i-2}{2n-1},\frac{2j-2}{2m-1}\right);$$

$$\Phi(f;X_2,Y_1;x,y) = \sum_{i=1}^{n}\sum_{j=1}^{m} \chi_i(A_2;x)\chi_j(A_1;y)f\left(\frac{i-1}{n-1},\frac{2j-2}{2m-1}\right);$$

$$\Phi(f;X_3,Y_1;x,y) = \sum_{i=1}^{n}\sum_{j=1}^{m} \chi_i(A_3;x)\chi_j(A_1;y)f\left(\frac{2i-1}{2n-1},\frac{2j-2}{2m-1}\right);$$

$$\Phi(f;X_0,Y_2;x,y) = \sum_{i=1}^{n}\sum_{j=1}^{m} \chi_i(A_0;x)\chi_j(A_2;y)f\left(\frac{2i-1}{2n},\frac{j-1}{m-1}\right);$$

(continued on p 98)

**Table 3.2.1**

| $\nu$ | $\mu$ | $n$ | $m$ | $\Phi(f; X_\nu, Y_\mu; x, y)$ | $H^{\omega_1\omega_2}(D_2)$ <br> $R^{\nu\mu}_{nm}(H^{\omega_1\omega_2}(D_2))_{C(D_2)}$ |
|---|---|---|---|---|---|
| 0 | 0 | | | $\Phi(f; X_0, Y_0; x, y)$ | $\omega_1\left(\dfrac{1}{2n}\right) + \omega_2\left(\dfrac{1}{2m}\right)$ |
| 1 | 0 | | | $\Phi(f; X_1, Y_0; x, y)$ | $\omega_1\left(\dfrac{1}{2n-1}\right) + \omega_2\left(\dfrac{1}{2m}\right)$ |
| 2 | 0 | $n>1$ | | $\Phi(f; X_2, Y_0; x, y)$ | $\omega_1\left(\dfrac{1}{2n-2}\right) + \omega_2\left(\dfrac{1}{2m}\right)$ |
| 3 | 0 | | | $\Phi(f; X_3, Y_0; x, y)$ | $\omega_1\left(\dfrac{1}{2n-1}\right) + \omega_2\left(\dfrac{1}{2m}\right)$ |
| 0 | 1 | | | $\Phi(f; X_0, Y_1; x, y)$ | $\omega_1\left(\dfrac{1}{2n}\right) + \omega_2\left(\dfrac{1}{2m-1}\right)$ |
| 1 | 1 | | | $\Phi(f; X_1, Y_1; x, y)$ | $\omega_1\left(\dfrac{1}{2n-1}\right) + \omega_2\left(\dfrac{1}{2m-1}\right)$ |
| 2 | 1 | $n>1$ | | $\Phi(f; X_2, Y_1; x, y)$ | $\omega_1\left(\dfrac{1}{2n-2}\right) + \omega_2\left(\dfrac{1}{2m-1}\right)$ |
| 3 | 1 | | | $\Phi(f; X_3, Y_1; x, y)$ | $\omega_1\left(\dfrac{1}{2n-1}\right) + \omega_2\left(\dfrac{1}{2m-1}\right)$ |
| 0 | 2 | | $m>1$ | $\Phi(f; X_0, Y_2; x, y)$ | $\omega_1\left(\dfrac{1}{2n}\right) + \omega_2\left(\dfrac{1}{2m-2}\right)$ |
| 1 | 2 | | $m>1$ | $\Phi(f; X_1, Y_2; x, y)$ | $\omega_1\left(\dfrac{1}{2n-1}\right) + \omega_2\left(\dfrac{1}{2m-2}\right)$ |
| 2 | 2 | $n>1$ | $m>1$ | $\Phi(f; X_2, Y_2; x, y)$ | $\omega_1\left(\dfrac{1}{2n-2}\right) + \omega_2\left(\dfrac{1}{2m-2}\right)$ |
| 3 | 2 | | $m>1$ | $\Phi(f; X_3, Y_2; x, y)$ | $\omega_1\left(\dfrac{1}{2n-1}\right) + \omega_2\left(\dfrac{1}{2m-2}\right)$ |
| 0 | 3 | | | $\Phi(f; X_0, Y_3; x, y)$ | $\omega_1\left(\dfrac{1}{2n}\right) + \omega_2\left(\dfrac{1}{2m-1}\right)$ |
| 1 | 3 | | | $\Phi(f; X_1, Y_3; x, y)$ | $\omega_1\left(\dfrac{1}{2n-1}\right) + \omega_2\left(\dfrac{1}{2m-1}\right)$ |
| 2 | 3 | $n>1$ | | $\Phi(f; X_2, Y_3; x, y)$ | $\omega_1\left(\dfrac{1}{2n-2}\right) + \omega_2\left(\dfrac{1}{2m-1}\right)$ |
| 3 | 3 | | | $\Phi(f; X_3, Y_3; x, y)$ | $\omega_1\left(\dfrac{1}{2n-1}\right) + \omega_2\left(\dfrac{1}{2m-1}\right)$ |

| $f_{12}(X_\nu, Y_\mu; x, y)$ | $H^\omega(D_2)$ | |
|---|---|---|
| | $R_{nm}^{\nu\mu}(H^\omega(D_2))_{C(D_2)}$ | $f_0(X_\nu, Y_\mu; x, y)$ |
| $f_{12}(X_0, Y_0; x, y)$ | $\omega\left(\dfrac{1}{2}\sqrt{\dfrac{1}{n^2} + \dfrac{1}{m^2}}\right)$ | $f_0(X_0, Y_0; x, y)$ |
| $f_{12}(X_1, Y_0; x, y)$ | $\omega\left(\dfrac{1}{2}\sqrt{\dfrac{1}{(n-1/2)^2} + \dfrac{1}{m^2}}\right)$ | $f_0(X_1, Y_0; x, y)$ |
| $f_{12}(X_2, Y_0; x, y)$ | $\omega\left(\dfrac{1}{2}\sqrt{\dfrac{1}{(n-1)^2} + \dfrac{1}{m^2}}\right)$ | $f_0(X_2, Y_0; x, y)$ |
| $f_{12}(X_3, Y_0; x, y)$ | $\omega\left(\dfrac{1}{2}\sqrt{\dfrac{1}{(n-1/2)^2} + \dfrac{1}{m^2}}\right)$ | $f_0(X_3, Y_0; x, y)$ |
| $f_{12}(X_0, Y_1; x, y)$ | $\omega\left(\dfrac{1}{2}\sqrt{\dfrac{1}{n^2} + \dfrac{1}{(m-1/2)^2}}\right)$ | $f_0(X_0, Y_1; x, y)$ |
| $f_{12}(X_1, Y_1; x, y)$ | $\omega\left(\dfrac{1}{2}\sqrt{\dfrac{1}{(n-1/2)^2} + \dfrac{1}{(m-1/2)^2}}\right)$ | $f_0(X_1, Y_1; x, y)$ |
| $f_{12}(X_2, Y_1; x, y)$ | $\omega\left(\dfrac{1}{2}\sqrt{\dfrac{1}{(n-1)^2} + \dfrac{1}{(m-1/2)^2}}\right)$ | $f_0(X_2, Y_1; x, y)$ |
| $f_{12}(X_3, Y_1; x, y)$ | $\omega\left(\dfrac{1}{2}\sqrt{\dfrac{1}{(n-1/2)^2} + \dfrac{1}{(m-1/2)^2}}\right)$ | $f_0(X_3, Y_1; x, y)$ |
| $f_{12}(X_0, Y_2; x, y)$ | $\omega\left(\dfrac{1}{2}\sqrt{\dfrac{1}{n^2} + \dfrac{1}{(m-1)^2}}\right)$ | $f_0(X_0, Y_2; x, y)$ |
| $f_{12}(X_1, Y_2; x, y)$ | $\omega\left(\dfrac{1}{2}\sqrt{\dfrac{1}{(n-1/2)^2} + \dfrac{1}{(m-1)^2}}\right)$ | $f_0(X_1, Y_2; x, y)$ |
| $f_{12}(X_2, Y_2; x, y)$ | $\omega\left(\dfrac{1}{2}\sqrt{\dfrac{1}{(n-1)^2} + \dfrac{1}{(m-1)^2}}\right)$ | $f_0(X_2, Y_2; x, y)$ |
| $f_{12}(X_3, Y_2; x, y)$ | $\omega\left(\dfrac{1}{2}\sqrt{\dfrac{1}{(n-1/2)^2} + \dfrac{1}{(m-1)^2}}\right)$ | $f_0(X_3, Y_2; x, y)$ |
| $f_{12}(X_0, Y_3; x, y)$ | $\omega\left(\dfrac{1}{2}\sqrt{\dfrac{1}{n^2} + \dfrac{1}{(m-1/2)^2}}\right)$ | $f_0(X_0, Y_3; x, y)$ |
| $f_{12}(X_1, Y_3; x, y)$ | $\omega\left(\dfrac{1}{2}\sqrt{\dfrac{1}{(n-1/2)^2} + \dfrac{1}{(m-1/2)^2}}\right)$ | $f_0(X_1, Y_3; x, y)$ |
| $f_{12}(X_2, Y_3; x, y)$ | $\omega\left(\dfrac{1}{2}\sqrt{\dfrac{1}{(n-1)^2} + \dfrac{1}{(m-1/2)^2}}\right)$ | $f_0(X_2, Y_3; x, y)$ |
| $f_{12}(X_3, Y_3; x, y)$ | $\omega\left(\dfrac{1}{2}\sqrt{\dfrac{1}{(n-1/2)^2} + \dfrac{1}{(m-1/2)^2}}\right)$ | $f_0(X_3, Y_3; x, y)$ |

$$\Phi(f;X_1,Y_2;x,y)=\sum_{i=1}^{n}\sum_{j=1}^{m}\chi_i(A_1;x)\chi_j(A_2;y)f\left(\frac{2i-2}{2n-1},\frac{j-1}{m-1}\right);$$

$$\Phi(f;X_2,Y_2;x,y)=\sum_{i=1}^{n}\sum_{j=1}^{m}\chi_i(A_2;x)\chi_j(A_2;y)f\left(\frac{i-1}{n-1},\frac{j-1}{m-1}\right);$$

$$\Phi(f;X_3,Y_2;x,y)=\sum_{i=1}^{n}\sum_{j=1}^{m}\chi_i(A_3;x)\chi_j(A_2;y)f\left(\frac{2i-1}{2n-1},\frac{j-1}{m-1}\right);$$

$$\Phi(f;X_0,Y_3;x,y)=\sum_{i=1}^{n}\sum_{j=1}^{m}\chi_i(A_0;x)\chi_j(A_3;y)f\left(\frac{2i-1}{2n},\frac{2j-1}{2m-1}\right);$$

$$\Phi(f;X_1,Y_3;x,y)=\sum_{i=1}^{n}\sum_{j=1}^{m}\chi_i(A_1;x)\chi_j(A_3;y)f\left(\frac{2i-2}{2n-1},\frac{2j-1}{2m-1}\right);$$

$$\Phi(f;X_2,Y_3;x,y)=\sum_{i=1}^{n}\sum_{j=1}^{m}\chi_i(A_2;x)\chi_j(A_3;y)f\left(\frac{i-1}{n-1},\frac{2j-1}{2m-1}\right);$$

$$\Phi(f;X_3,Y_3;x,y)=\sum_{i=1}^{n}\sum_{j=1}^{m}\chi_i(A_3;x)\chi_j(A_3;y)f\left(\frac{2i-1}{2n-1},\frac{2j-1}{2m-1}\right).$$

# 4

---

# Some quadrature formulas

This chapter is devoted to optimal (in one or other sense) quadrature formulas for functions of the classes $W^r H^\omega$ ($r = 0, 2, 4, \ldots$) with an arbitrary modulus of continuity $\omega(t)$. This is done using only the information $T_r(f, X)$ (see (1.1.4)) for the function $f$ on the mesh $X = (x_1, x_2, \ldots, x_n)$ both in free nodes, as in nodes subjected to an additional condition.

## 4.1   The best quadrature formula for functions of the class $H^\omega$

The best (optimal) quadrature formula for the integral

$$Lf = \int_0^1 f(x)\mathrm{d}x \tag{4.1.1}$$

in the class $W^r H^\omega$ on information $T_r(f, X)$ is said to be the best (optimal) method for recovery of the functional in the class $W^r H^\omega$ on the information $T_r(f, X)$.

**Lemma 4.1.1**

*Let $n \in \mathbb{N}$, $X \in (X)_0$ and $\omega(t)$ $(0 \leqq t \leqq 1)$ be an arbitrary modulus of continuity. Then the identity*

$$\sup\{\int_0^1 f(x)\mathrm{d}x : f \in H_0^\omega\} = L_n(\omega; B, X) \tag{4.1.2}$$

*with $H_o^\omega$ defined in (2.2.2), $L_n(\omega; B, X)$ as in (2.1.2), and $B = (b_0, b_1, \ldots, b_n)$ as in (1.5.21) holds.*

## Proof

According to Lemma 2.2.1, for every function $f \in H_0^\omega$ the inequality

$$f(x) \leqq f_0(\omega, X; x), \quad x \in [0, 1]$$

holds and therefore, the inequality

$$\int_0^1 f(x)\mathrm{d}x \leqq \int_0^1 f_0(\omega, X; x)\mathrm{d}x$$

is satisfied.

Using the fact that the standard function $f_0(\omega, X, x)$ also belongs to the class $H_0^\omega$ (see Property (v) of Lemma 2.2.1) we obtain the equality[†]

$$\sup\left\{\int_0^1 f(x)\mathrm{d}x : f \in H_0^\omega\right\} = \int_0^1 f_0(\omega, X; x)\mathrm{d}x. \tag{4.1.3}$$

From Definition (2.2.1) of the standard function $f_0(\omega, X; x)$ and from Definition (2.2.1) of $L_n(\varphi; A, X)$ the equality

$$\int_0^1 f_0(\omega, X; x)\mathrm{d}x = \sum_{i=1}^n \int_{b_{i-1}}^{b_i} \omega(|x - x_i|)\mathrm{d}x = L_n(\omega; B, X) \tag{4.1.4}$$

follows. This equality together with (4.1.3) gives (4.1.2) and thus the lemma is proved.

**Corollary 4.1.1** *Under the hypothesis of Lemma 4.1.1 the inequality*

$$\sup\left\{\int_0^1 f(x)\mathrm{d}x : f \in H_0^\omega\right\} \leqq \omega(1) \tag{4.1.5}$$

*holds.*

*Indeed, from (4.1.4), (1.6.14) and the monotonicity of $\omega(t)$ the inequality*

$$L_n(\omega; B, X) \leqq \omega(D(B, X)_\infty) \leqq \omega(1)$$

*follows.*

*From this inequality and from (4.1.2) follows (4.1.5).*

[†] It may be written in the form $\sup\limits_{f \in H_0^\omega} \int_0^1 f(x)\mathrm{d}x = \int_0^1 \sup\limits_{f \in H_0^\omega} f(x)\mathrm{d}x.$

## Lemma 4.1.2

*Let $n \in \mathbb{N}$, $X \in (X)_0$ and $\omega(t)$ be an arbitrary modulus of continuity. Then the error of the best quadrature formula for the integral (4.1.1) in the function class $H^\omega$ on the information $T_0(f, X)$, $X \in (X)_0$ is determined by*

$$R(H^\omega, X) = L_n(\omega; B, X) \tag{4.1.6}$$

*with the same notation as in Lemma 4.1.1.*

## Proof

Since the linear functionals $Lf = \int_0^1 f(x)\mathrm{d}x$, $L_i f = f(x_i)$, $i = 1, 2, \ldots, n$ are defined in the convex centrally symmetric body $H^\omega$ of the normed linear space $C[0, 1]$ and the inequality (4.1.5) holds, then all the conditions for the validity of Smolyak's lemma are satisfied.

Let us denote by $S$ the set of all possible quadrature formulas for the integral (4.1.1) in the class $H^\omega$ using the information $T_0(f, X)$ only, then, for the error

$$R(H^\omega, X) = \inf_{Q \in S} \sup_{f \in H^\omega} |Lf - Q|$$

of the best quadrature formula for (4.1.1) in $H^\omega$ on $T_0(f, X)$ according to the corollary of Smolyak's lemma, the equation

$$R(H^\omega, X) = \sup\left\{ \int_0^1 f(x)\mathrm{d}x : f \in H_0^\omega \right\}$$

holds. From it and (4.1.2) follows (4.1.6) and thus the lemma is proved.

Let us write

$$\int_0^1 f(x)\mathrm{d}x = Q^0(f, X) + R(f; X) \tag{4.1.7}$$

with

$$Q^0(f; X) = \int_0^1 \Phi^0(f; X; x)\mathrm{d}x = \sum_{i=1}^n \int_{b_{i-1}}^{b_i} f(x_i)\mathrm{d}x = \sum_{i=1}^n (b_i - b_{i-1})f(x_i) \tag{4.1.8}$$

where $\Phi^0(f; X; x)$ is the best method for recovery of functions of the class $H^\omega$ on information $T_0(f, X)$, defined in (2.3.3).

### Theorem 4.1.1

*Let $n \in \mathbb{N}$, $X \in (X)_0$ be an arbitrary mesh in the interval $[0, 1]$ and let $\omega(t)$ $(0 \leq t \leq 1)$ be an arbitrary modulus of continuity. Then out of all quadrature formulas for recovery of integral (4.1.1) in the class $H^\omega$ on information $T_0(f, X)$ the best is formula (4.1.7), (4.1.8) with the error*

$$R(H^\omega, X) = \sup_{f \in H^\omega} \left| \int_0^1 f(x)\mathrm{d}x - Q^0(f, X) \right| = L_n(\omega; B, X). \qquad (4.1.9)$$

*The supremum in (4.1.9) is attained by the standard function $f_0(\omega, X; x)$ determined in (2.2.1).*

### Proof

For each function $f \in H^\omega$ and for every mesh $X \in (X)_0$ from (4.1.7) and (4.1.8) it follows that

$$|R(f; X)| = \left| \int_0^1 f(x)\mathrm{d}x - Q^0(f, X) \right|$$

$$= \left| \sum_{i=1}^n \int_{b_{i-1}}^{b_i} [f(x) - f(x_i)]\mathrm{d}x \right|$$

$$\leq \sum_{i=1}^n \int_{b_{i-1}}^{b_i} |f(x) - f(x_i)|\mathrm{d}x \leq \sum_{i=1}^n \int_{b_{i-1}}^{b_i} \omega(|x - x_i|)\mathrm{d}x$$

$$= L_n(\omega; B, X) \overset{(4.1.4)}{=} \int_0^1 f_0(\omega, X; x)\mathrm{d}x$$

that is

$$\sup_{f \in H^\omega} |R(f, X)| \leq L_n(\omega; B, X) = \int_0^1 f_0(\omega, X; x)dx. \tag{4.1.10}$$

On the other hand, from the properties $f_0(\omega, X; x) \geq 0, f_0 \in H_0^\omega$ and equation (2.3.5) the following equation for the standard function $f_0(\omega, X; x)$ follows:

$$R(f_0; X) = \int_0^1 f_0(\omega, X; x)dx - Q^0(f_0; X) = \int_0^1 f_0(\omega, X; x)dx. \tag{4.1.11}$$

From (4.1.10) and (4.1.11), since $f_0 \in H^\omega$ then it follows

$$\sup_{f \in H^\omega} |R(f; X)| = L_n(\omega; B, X) = \int_0^1 f_0(\omega, X; x)dx \tag{4.1.12}$$

and from (4.1.12) and (4.1.6) follows (4.1.9). By definition (1.5.3) and from (4.1.9) it follows that quadrature formula (4.1.7), (4.1.8) is the best for integral (4.1.1) in the class $H^\omega$ on information $T_0(f; X)$. Then (4.1.11) implies that the supremum in (4.1.9) is attained by the standard function $f_0(\omega, X; x)$. The theorem is proved.

## 4.2   Optimal quadrature formulas for functions of the class $H^\omega$

### Theorem 4.2.1

*Let $n \in \mathbb{N}$, $v$ be a fixed element of the set $\{0, 1, 2, 3\}$, $X \in (X)$ be a mesh on the segment $[0, 1]$ and $\omega(t)$ $(0 \leq t \leq 1)$ be an arbitrary modulus of continuity. Then, out of all quadrature formulas for recovery of integral (4.1.1) in the class $H^\omega$ on information $T_0(f, X)$, $X \in (X)_v$ optimal is the formula*

$$\int_0^1 f(x)dx = Q^0(f, X_v) + R(f, X_v) \tag{4.2.1}$$

*where $Q^0(f, X)$ is determined by (4.1.8), and $X_v \in (X)_v$ by (1.6.5$_v$) respectively.*

*Furthermore, for $v = 0, 1, 2, 3$ we have*

$$R_n^v(H^\omega) = \inf_{X \in (X)_v} R(H^\omega, X) = R(H^\omega, X_v) = \int_0^1 f_0(\omega, X_v; x) \mathrm{d}x$$

$$= \sup_{f \in H^\omega} \left| \int_0^1 f(x)\mathrm{d}x - Q^0(f, X_v) \right| = R(f_0, X_v) = \int_0^1 \omega(u \cdot D(A_v, X_v)_\infty) \mathrm{d}u$$

$$(4.2.2)$$

*where $f_0(\omega, X_v; x)$ is the standard function (2.2.1), the mesh pair $(A_v, X_v)$ is determined by $(1.6.5_v)$, and $D(A, X)_\infty$ is the uniform deviation of the mesh pair $(A, X)$.*

*From (4.2.2) it follows that the standard function $f_0(\omega, X_v; x)$ is an extremal for (4.2.1) in $H^\omega$ on $T_0(f, X)$, $X \in (X)_v$.*

## Proof

Let $f \in H^\omega$. By definition (see (1.5.15))

$$R_n^v(H^\omega) = \inf\{R(H^\omega, X): X \in (X)_v\}.$$

By Theorem 4.1.1 (see (4.1.9)), we have

$$\inf_{X \in (X)_v} R(H^\omega, X) = \inf_{X \in (X)_v} L_n(\omega; B, X) \geqq \inf_{(A,X) \in (A,X)_v} L_n(\omega; A, X).$$

According to the optimization theorem (see (2.1.11))

$$\inf_{(A,X) \in (A,X)_v} L_n(\omega; A, X) = L_n(\omega; A_v. X_v) \overset{(4.1.9)}{=} R(H^\omega, X_v)$$

since, as can be easily verified, the mesh $A_v$ is determined uniquely by the mesh $X_v$ by formulas (1.5.21).

From the last three relations it follows that

$$R_n^v(H^\omega) = \inf\{R(H^\omega, X): X \in (X)_v\} \geqq R(H^\omega, X_v) = L_n(\omega; A_v, X_v).$$

Since $X_v \in (X)_v$, then we get

$$R_n^v(H^\omega) = \inf\{R(H^\omega, X): X \in (X)_v\} = R(H^\omega, X_v) = L_n(\omega; A_v, X_v). \quad (4.2.3)$$

From (4.2.3) and (2.1.3) it follows that

$$R_n^v(H^\omega) = L_n(\omega; A_v, X_v) = \int_0^1 \omega(u \cdot D(A_v, X_v)_\infty) \, du. \tag{4.2.4}$$

From (4.1.9) we get

$$R(H^\omega, X_v) = \sup_{f \in H_\omega} \left| \int_0^1 f(x)dx - Q^0(f, X_v) \right| \tag{4.2.5}$$

and from (4.1.11) and (4.1.12)

$$L_n(\omega; A_v, X_v) = \int_0^1 f_0(\omega, X_v; x)dx = \int_0^1 f_0(\omega, X_v; x)dx - Q^0(f_0, X_v). \tag{4.2.6}$$

Relations (4.2.3)–(4.2.6) imply (4.2.2) and thus all the assertions of the theorem since, by definition, (4.2.3) means the optimality of the mesh $X_v$ and hence the optimality of $Q^0(f, X_v)$. Relations (4.2.6) mean that the standard function $f_0(\omega, X_v; x)$ is an extremal. Thus the theorem is proved.

This theorem for $v = 0$ implies the validity of the assertion.

**Corollary 4.2.1** *(Korneichuck [19].) Let $n \in \mathbb{N}$ and $\omega(t)$ $(0 \le t \le 1)$ be an arbitrary modulus of continuity. Then out of all possible quadrature formulas for recovery of integral (4.1.1) in the class $H^\omega$ on information $T_0(f, X)$, $X \in (X)_v$ the optimal is the quadrature formula of the midpoint*

$$\int_0^1 f(x)dx = 1/n \sum_{i=1}^n f((2i-1)/2n) + R(f, X_0) \tag{4.2.7}$$

*with the error*

$$R_n^0(H^\omega) = \int_0^1 \omega(u/2n)du. \tag{4.2.8}$$

From corollary 4.1.2, for $\omega(t) = t^\alpha, 0 < \alpha \le 1$, we get the following corollary.

**Corollary 4.2.2** (*Turetzkii* [*18*].) *Let* $n \in \mathbb{N}$, $0 < \alpha \le 1$ $\omega(t) = t^{\alpha}$, $0 \le t \le 1$. *Then, out of all possible quadrature formulas for recovery of the integral* (*4.1.1*) *in the class* $H^{\omega}$ *on information* $T_0(f, X)$, $X \in (X)_0$ *the optimal one is* (*4.2.7*) *of the midpoint with the error* (*1.5.19*).

Theorem 4.2.1 for $v = 1$ implies

**Corollary 4.2.3** *Let* $n \in \mathbb{N}$ *and* $\omega(t)$ $(0 \le t \le 1)$ *be an arbitrary modulus of continuity. Then, out of all possible quadrature formulas for recovery of integral* (*4.1.1*) *in the class* $H^{\omega}$ *on information* $T_0(f, X)$, $X \in (X)_1$, *the optimal one is*

$$\int_0^1 f(x)dx = \frac{2}{2n-1}\left[\frac{f(0)}{2} + \sum_{i=2}^{n} f\left(\frac{2i-2}{2n-1}\right)\right] + R(f, X_1) \qquad (4.2.9)$$

*with the error*

$$R_n^1(H^{\omega}) = \int_0^1 \omega(u/(2n-1))du. \qquad (4.2.10)$$

*In this case the standard function* $f_0(\omega, X_1; x)$ *is extremal.*
*In the special case* $\omega(t) = t^{\alpha}$, $0 \le t \le 1$ *we have*

$$R_n^1(H^{\alpha}) = 1/(\alpha + 1)(2n - 1)^{\alpha}. \qquad (4.2.11)$$

*Since* $H^1 = W_{\infty}^1$ *for* $\alpha = 1$ *then*

$$R_n^1(W_{\infty}^1) = 1/2(2n - 1). \qquad (4.2.12)$$

Following Nikol'skii [1], p 26, let us denote $W_0^1[1; 0, 1] = \{f(x): f \in W_{\infty}^1, f(0) = 0\}$. Then, Corollary 4.2.3 implies the following corollary.

**Corollary 4.2.4** ([*1*], *p 90*.) *Let* $n = m + 1$, $m \in \mathbb{N}$. *Out of all quadrature formulas for recovery of the integral* (*4.1.1*) *in the class* $W_0^1[1; 0, 1]$ *on information* $T_0(f, X), X \in (X)_1$ *the optimal one is the formula*

$$\int_0^1 f(x)dx = \frac{2}{2m+1} \sum_{i=2}^{m+1} f\left(\frac{2i-2}{2m+1}\right) + R(f, X_1) \qquad (4.2.13)$$

*with the error*

$$R^1_{m+1}(W^1_0[1;0,1]) = 1/2(2m+1). \qquad (4.2.14)$$

From Theorem 4.2.1 for $v = 2$ follows Corollary 4.2.5.

**Corollary 4.2.5**  *Let $n = 2,3,\ldots$, and $\omega(t)$ be an arbitrary modulus of continuity. Then, out of all quadrature formulas for approximate evaluation of the integral (4.1.1) in the function class $H^\omega$ on information $T_0(f,X)$, $X \in (X)_2$, optimal is the trapezoidal formula*

$$\int_0^1 f(x)\mathrm{d}x = \frac{1}{n-1}\left[\frac{f(0)+f(1)}{2} + \sum_{i=2}^{n-1} f\left(\frac{i-1}{n-1}\right)\right] + R(f,X_2) \qquad (4.2.15)$$

*with the error*

$$R^2_n(H^\omega) = \int_0^1 \omega(u/(2n-2))\mathrm{d}u. \qquad (4.2.16)$$

*In the special case $\omega(t) = t^\alpha$, $0 < \alpha \leq 1$ we have*

$$R^2_n(H^\alpha) = 1/(\alpha+1)(2n-2)^\alpha$$

*and for $\alpha = 1$*

$$R^2_n(H^1) = R^2_n(W^1_\infty) = 1/4(n-1)$$

*respectively, i.e. the known classical estimation.*

Finally, from Theorem 4.2.6 for $v = 3$ we obtain the following corollary.

**Corollary 4.2.6**  *Let $n \in \mathbb{N}$, and $\omega(t)$, $(0 \leq t \leq 1)$ be an arbitrary modulus of continuity. Then, out of all quadrature formulas for recovery of the integral (4.1.1) in the class $H^\omega$ on information $T_0(f,X)$, $X \in (X)_3$, optimal is the formula*

$$\int_0^1 f(x)\mathrm{d}x = \frac{2}{2n-1}\left[\frac{f(1)}{2} + \sum_{i=1}^{n-1} f\left(\frac{2i-1}{2n-1}\right)\right] + R(f,X_3) \qquad (4.2.17)$$

*with the error*

$$R_n^3(H^\omega) = \int_0^1 \omega(u/(2n-1)\mathrm{d}u. \qquad (4.2.18)$$

*In the special case $\omega(t) = t^\alpha$, $0 < \alpha \leqq 1$, we get*

$$R_n^3(H^\alpha) = 1/(\alpha + 1)(2n-1)^\alpha$$

*and for $\alpha = 1$*

$$R_n^3(H^1) = R_n^3(W_\infty^1) = 1/2(2n-1)$$

*respectively.*

## 4.3   An estimation for the error of optimal quadrature formulas for individual functions in $H^\omega$

In this section we will highlight the 'ends effect', i.e. the influence of restrictions $x_1 = 0$ and $x_n = 1$ or $x_1 = 0$ and $x_n = 1$ on the errors of the optimal quadrature formulas, considered in Section 4.2, as far as it concerns individual functions.

**Proposition 4.3.1**   *For every positive integer n and for every Riemann integrable function $f : [0, 1] \to \mathbb{R}$ the estimate*

$$|R(f, X_0)| = \left| \int_0^1 f(x)\mathrm{d}x - \frac{1}{n} \sum_{i=1}^n f\left(\frac{2i-1}{2n}\right) \right| \leqq \frac{1}{2} \int_0^1 \omega_2(f; u/2n)\mathrm{d}u \quad (4.3.1)$$

*holds, where $\omega_2(f; \delta)$ is the second modulus of continuity (Zigmund's modulus) of the function f.*

### Proof

Let $n \in \mathbb{N}$ and $f \in R[0, 1]$. Then

$$|R(f, X_0)| = \left| \int_0^1 f(x)\mathrm{d}x - \frac{1}{n} \sum_{i=1}^n f\left(\frac{2i-1}{2n}\right) \right|$$

$$= \left| \sum_{i=1}^n \int_{(i-1)/n}^{i/n} \left[ f(x) - f\left(\frac{2i-1}{2n}\right) \right] \mathrm{d}x \right|$$

$$= \left| \sum_{i=1}^{n} \left\{ \int_{(i-1)/n}^{(2i-1)/2n} \left[ f(x) - f\left(\frac{2i-1}{2n}\right) \right] dx \right.\right.$$

$$\left.\left. + \int_{(2i-1)/2n}^{i/n} \left[ f(x) - f\left(\frac{2i-1}{2n}\right) dx \right\} \right| \right.$$

$$= \frac{1}{2n} \left| \sum_{i=1}^{n} \int_{0}^{1} \left[ f\left(\frac{2i-1}{2n} + \frac{u}{2n}\right) - 2f\left(\frac{2i-1}{2n}\right) + f\left(\frac{2i-1}{2n} - \frac{u}{2n}\right) \right] du \right|$$

$$\le \frac{1}{2n} \sum_{i=1}^{n} \int_{0}^{1} \left| f\left(\frac{2i-1}{2n} + \frac{u}{2n}\right) - 2f\left(\frac{2i-1}{2n}\right) + f\left(\frac{2i-1}{2n} - \frac{u}{2n}\right) \right| du$$

$$\le \frac{1}{2} \int_{0}^{1} \omega_2\left(f; \frac{u}{2n}\right) du$$

since according to Definition (1.2.3) of $\omega_2(f;\delta)$ we have

$$|f(x+\delta) - 2f(x) + f(x-\delta)| \le \omega_2(f;\delta), \quad x \pm \delta \in [0,1].$$

The proof is completed.

For each $\alpha \in (0,2]$ we denote

$$Z_M^\alpha = \{f(x): \omega_2(f;t) \le Mt^\alpha, \, M > 0, \, 0 \le t \le \tfrac{1}{2}\}.$$

For $\alpha = 1$ we will write $Z_M$ instead of $Z_M^1$.

**Corollary 4.3.1**  *For every $n \in \mathbb{N}$ and for each function $f \in Z_M^\alpha$, $0 < \alpha \le 2$ the inequality*

$$\left| \int_{0}^{1} f(x)dx - \frac{1}{n} \sum_{i=1}^{n} f\left(\frac{2i-1}{2n}\right) \right| \le \frac{M}{2^{\alpha+1}(\alpha+1)n^\alpha} \tag{4.3.2}$$

*is satisfied.*

For functions of the Zigmund class (i.e. for $\alpha = 1$) we obtain from (4.3.2) the estimate

$$\left| \int_0^1 f(x)dx - \frac{1}{n} \sum_{i=1}^n f\left(\frac{2i-1}{2n}\right) \right| \leq \frac{M}{8n} \qquad (4.3.3)$$

*and for $\alpha = 2$ we obtain the estimate*

$$\left| \int_0^1 f(x)dx - \frac{1}{n} \sum_{i=1}^n f\left(\frac{2i-1}{2n}\right) \right| \leq \frac{M}{24n^2} \qquad (4.3.4)$$

*known from the calculus.*

**Corollary 4.3.2** *Let $n \in \mathbb{N}$. Then for each Riemann integrable function $f:[0,1] \to \mathbb{R}$ the inequality*

$$\left| \int_0^1 f(x)dx - \frac{1}{n} \sum_{i=1}^n f\left(\frac{2i-1}{2n}\right) \right| \leq \frac{1}{2} \int_0^1 \omega_2\left(f;\frac{u}{2n}\right)du \leq \int_0^1 \omega\left(f;\frac{u}{2n}\right)du \quad (4.3.5)$$

*holds. In the special case of a function $f$ of the class $H^\omega$ with an arbitrary modulus of continuity $\omega(t)$ the corresponding inequality*

$$\left| \int_0^1 f(x)dx - \frac{1}{n} \sum_{i=1}^n f\left(\frac{2i-1}{2n}\right) \right|$$

$$\leq \frac{1}{2} \int_0^1 \omega_2\left(f;\frac{u}{2n}\right)du \leq \int_0^1 \omega\left(f;\frac{u}{2n}\right)du \leq \int_0^1 \omega\left(\frac{u}{2n}\right)du \quad (4.3.6)$$

*holds.*

Indeed, inequalities (4.3.5) and (4.3.6) follow from (4.3.1) and from the known properties of the Zygmund modulus for the function $f \in H^\omega$ (see Section 1.2) as follows:

$$\omega_2(f;\delta) \leq 2\omega(f;\delta) \leq 2\omega(\delta).$$

Since, according to Theorem 4.2.1, inequality (4.3.6) is exact, then (4.3.1) is also an exact inequality.

**Corollary 4.3.3**   *Let $n \in \mathbb{N}$. Then for each function $f \in C^1$ the inequality*

$$\left| \int_0^1 f(x)\mathrm{d}x - \frac{1}{n} \sum_{i=1}^n f\left(\frac{2i-1}{2n}\right) \right| \leq \frac{1}{4n} \int_0^1 u \cdot \omega\left(f'; \frac{u}{2n}\right) \mathrm{d}u \qquad (4.3.7)$$

*is satisfied.*

*In the special case $f' \in MH^\alpha$, $0 < \alpha \leq 1$, the corresponding inequality has the form*

$$\left| \int_0^1 f(x)\mathrm{d}x - \frac{1}{n} \sum_{i=1}^n f\left(\frac{2i-1}{2n}\right) \right| \leq \frac{M}{2^{2+\alpha}(2+\alpha)n^{1+\alpha}}. \qquad (4.3.8)$$

Indeed, inequality (4.3.7) follows from (4.3.1) and from the known property

$$\omega_2(f; \delta) \leq \delta \cdot \omega(f'; \delta)$$

of the Zygmund modulus (see Section 1.2). As for inequality (4.3.8), it follows from (4.3.7) and

$$f' \in MH^\alpha \Leftrightarrow \omega(f'; \delta) \leq M\delta^\alpha.$$

**Proposition 4.3.2**   *For every $n = 2, 3, \ldots$ and for each Riemann integrable function $f : [0, 1] \to \mathbb{R}$ the estimate*

$$\left| \int_0^1 f(x)\mathrm{d}x - \frac{1}{n-1} \left[ \frac{f(0)+f(1)}{2} + \sum_{i=2}^{n-1} f\left(\frac{i-1}{n-1}\right) \right] \right| \leq \int_0^1 \omega\left(f; \frac{u}{2n-2}\right) \mathrm{d}u$$

$$(4.3.9)$$

*holds.*

Indeed,

$$R(f, X_2) = \int_0^1 f(x)dx - \frac{1}{n-1}\left[\frac{f(0)+f(1)}{2} + \sum_{i=2}^{n-1} f\left(\frac{i-1}{n-1}\right)\right]$$

$$= \sum_{i=1}^{n-1}\left\{\int_{\frac{i-1}{n-1}}^{\frac{2i-1}{2n-2}}\left[f(x)-f\left(\frac{i-1}{n-1}\right)\right]dx + \int_{\frac{2i-1}{2n-2}}^{\frac{i}{n-1}}\left[f(x)-f\left(\frac{i}{n-1}\right)\right]dx.$$

By means of the substitution

$$x = \frac{i-1}{n-1} + \frac{u}{2n-2}$$

in the first of the integrals, and by the substitution

$$x = \frac{i}{n-1} - \frac{u}{2n-2}$$

in the second, we obtain

$$R(f, X_2) = \frac{1}{2n-2}\sum_{i=1}^{n-1}\left\{\int_0^1\left[f\left(\frac{i-1}{n-1}+\frac{u}{2n-2}\right)-f\left(\frac{i-1}{n-1}\right)\right]du\right.$$

$$\left. + \int_0^1\left[f\left(\frac{i}{n-1}-\frac{u}{2n-2}\right)-f\left(\frac{i}{n-1}\right)\right]du\right\}. \qquad (4.3.10)$$

From this and from the definition of the modulus of continuity of the function $f$ it follows that

$$|R(f, X_2)| \le \frac{1}{2n-2}\sum_{i=1}^{n-1} 2\cdot\int_0^1 \omega\left(f;\frac{u}{2n-2}\right)du = \int_0^1 \omega\left(f;\frac{u}{2n-2}\right)du.$$

Thus the proof is completed.

**Proposition 4.3.3**   *From evry $n = 2, 3, \ldots$ and for each function $f \in C'[0, 1]$ the inequality*

$$\left| \int_0^1 f(x)\,dx - \frac{1}{n-1}\left[ \frac{f(0)+f(1)}{2} + \sum_{i=2}^{n-1} f\!\left(\frac{i-1}{n-1}\right)\right]\right|$$

$$\leqq \frac{1}{4(n-1)} \int_0^1\!\!\int_0^1 u\cdot\omega\!\left( f'; \frac{1-us}{n-1}\right)du\,ds \qquad (4.3.11)$$

*is satisfied.*

Indeed, from (4.3.10) and from $f \in C^1[0,1]$ it follows that

$$R(f;X_2) = \frac{1}{2n-2}\sum_{i=1}^{n-1}\int_0^1\left[ \int_{\frac{i-1}{n-1}}^{\frac{i-1}{n-1}+\frac{u}{2n-2}} f'(t)\,dt + \int_{\frac{i}{n-1}}^{\frac{i}{n-1}-\frac{u}{2n-2}} f'(t)\,dt\right]du.$$

Making the substitutions

$$t = \frac{i-1}{n-1} + \frac{us}{2n-2}$$

and

$$t = \frac{i}{n-1} - \frac{us}{2n-2}$$

in the first and the second integral, respectively, we get

$$R(f,X_2) = \frac{i}{4(n-1)^2}\sum_{i=1}^{n-1}\int_0^1\!\!\int_0^1\left[ f'\!\left(\frac{i-1}{n-1} + \frac{us}{2n-2}\right)\right.$$

$$\left. - f'\!\left(\frac{1}{n-1} - \frac{us}{2n-2}\right)\right]u\,ds\,du.$$

This relation, along with the definition of the modulus of continuity of $f'$, imply the estimate

$$|R(f,X_2)| \leqq \frac{1}{4(n-1)}\cdot\int_0^1\!\!\int_0^1 u\cdot\omega\!\left( f'; \frac{1-us}{n-1}\right)ds\,du.$$

Thus (4.3.11) is proved.

**Corollary 4.3.4** *If* $n = 2, 3, \ldots$, *and* $f \in H^\omega$ *with an arbitrary modulus of continuity* $\omega(t)$, *then*

$$|R(f, X_2)| \leqq \int_0^1 \omega(u/(2n-2))\mathrm{d}u. \tag{4.3.12}$$

Since, as we already know, inequality (4.3.12) is exact, then inequality (4.3.9) is exact too. It is reduced to an identity for the standard function $f_0(\omega, X_2; x)$.

---

**Remark.** Comparing (4.3.5) and (4.3.9), we see that by fixing the nodes $x_1 = 0$ and $x_n = 1$, the estimate of the quadrature formula should be changed from

$$\int_0^1 \omega(f; u/2n)\mathrm{d}u \quad \text{to} \quad \int_0^1 \omega(f; u/(2n-2))\mathrm{d}u$$

for each individual function $f \in R[0, 1]$.

---

**Corollary 4.3.5** *If* $n = 2, 3, \ldots$, *and* $f' \in MH^\alpha$, $0 < \alpha \leq 1$, *then*

$$|R(f, X_2)| \leqq M/4(\alpha + 2)(n-1)^{1+\alpha}. \tag{4.3.13}$$

In the special case $\alpha = 1$, (4.3.13) implies the estimate

$$|R(f, X_2)| \leqq M/12(n-1)^2 \tag{4.3.14}$$

known from the calculus.

Comparing (4.3.8) with (4.3.13) and (4.3.4) with (4.3.14), we see the 'effect of the ends' in the corresponding cases.

A similar effect appears in the optimal formulas in $H^\omega$ with a fixed left or right node.


## 4.4 Optimization of the quadrature formula of Niederreiter type for the function classes $W^r H^\omega$

Niederreiter [31] has shown that for each mesh $X \in (X)_0$ and for every function $f \in C$ then

$$\left| \int_0^1 f(x)dx - \frac{1}{n} \sum_{i=1}^n f(x_i) \right| \leq \omega(f; D(A_0, X)_\infty) \tag{4.4.1}$$

holds, with $A_0 = (0, 1/n, 2/n, \ldots, 1)$.

Proinov [20] generalized Niederreiter by showing that for $X \in (X)_0$ and $f \in C$

$$\left| \int_0^1 f(x)dx - \sum_{i=1}^n p_i f(x_i) \right| \leq \omega(f; D(A, X)_\infty)$$

where

$$p_i \geq 0, \qquad \sum_{i=1}^n p_i = 1, \qquad a_0 = 0, \qquad a_i = \sum_{k=1}^i p_k, \qquad A = (a_0, a_1, \ldots, a_n).$$

In the papers of Proinov and Kirov [32], [33], Christov [34] and Totkov and Bazelkov [35], quadrature formulas generalizing Niederreiter's result for some function classes, having $r$th ($r \in \mathbb{N}$) derivatives, are constructed.

Here we consider the optimization problem of the quadrature formula of Niederreiter type for the function classes $W^r H^\omega$. In this section we give proofs (in elaborate and extended form) of results announced in [26].

So, let $n \in \mathbb{N}$, $A = (a_0, a_1, \ldots, a_n)$ and $X = (x_1, x_2, \ldots, x_n)$; let $(A, X) \in (A, X)_0$ be two arbitrary meshes in the segment $[0, 1]$, $\omega(t)$ ($0 \leq t \leq 1$) be an arbitrary modulus of continuity, $r \in \mathbb{N}$, and $f \in W^r H^\omega$.

The quadrature formula

$$\int_0^1 f(x)dx = Q^r(f; A, X) + R_r(f; A, X) \tag{4.4.2}$$

where

$$Q^r(f; A, X) = \sum_{i=q}^n \sum_{k=0}^r \lambda_{ik} f_{(x_i)}^{(k)} \tag{4.4.3}$$

and

$$\lambda_{ik} = \frac{(a_i - x_i)^{k+1} - (a_{i-1} - x_i)^{k+1}}{(k+1)!} \qquad (i = 1, 2, \ldots, n; k = 0, 1, \ldots, r) \tag{4.4.4}$$

is said to be a quadrature formula of Niederreiter type.

By means of a given mesh $X = (x_1, x_2, \ldots, x_n)$ we build a mesh $B = (b_0, b_1, \ldots, b_n)$ uniquely determined by the formulas

$$b_0 = 0, \quad b_n = 1, \quad b_i = \frac{x_i + x_{i+1}}{2} \quad (i = 1, 2, \ldots, n-1). \tag{4.4.5}$$

Further we build the quadrature formula

$$\int_0^1 f(x)\,dx = Q^r(f; B, X) + R_r(f; B, X) \tag{4.4.6}$$

on the basis of the mesh pair $(B, X) \in (A, X)_0$ by formulas (4.4.2)–(4.4.4). It has the properties:
 (i) uses *only* the information $T_r(f; X)$;
 (i) holds the equality

$$Q^r(f; B, X) = \int_0^1 \Phi^r(f, X; x)\,dx \tag{4.4.7}$$

where $\Phi^r(f, X; x)$ is the quasi-spline of degree $r$, defined in (1.5.25);
(iii) since the mesh $B$ is uniquely determined by $X$, then $R_r(f; B, X) = R_r(f; X)$.

### Lemma 4.3.1

*Let $n \in \mathbb{N}$, $r(r \in \mathbb{N})$ be an even number, $X \in (X)_0$ and $\omega(t)$ $(0 \leq t \leq 1)$ be an arbitrary modulus of continuity. Then*

$$\Delta(W^r H^\omega; X) \overset{\text{def}}{=} \sup\{|R_r(f; B, X)|: f \in W^r H^\omega\} = R_r(f_r; B, X) = L_n(\varphi; B, X) \tag{4.4.8}$$

*where $L_n(\varphi; A, X)$ is determined by (2.1.2)*

$$\varphi(u) = \frac{u^r}{(r-1)!} \int_0^1 (1-z)^{r-1}\omega(zu)\,dz, \quad u \in [0, 1] \tag{4.4.9}$$

*and $f_r(x) = f_r(\omega, X; x)$ is the function determined in (2.5.2).*

*Additionally, if for a mesh pair $(A, X) \in (A, X)_0$ the equation*

$$\sup\{|R_r(f; A, X)|: f \in W^r H^\omega\} = L_n(\varphi; A, X) \qquad (4.4.10)$$

*is satisfied, then the inequality*

$$\sup_{f \in W^r H^\omega} |R_r(f; A, X)| \geqq \sup_{f \in W^r H^\omega} |R_r(f; B, X)| \qquad (4.4.11)$$

*holds.*

The meaning of this assertion is that

$$\sup_{f \in W^r H^\omega} |R_r(f; B, X)| = L_n(\varphi; B, X) = \inf_A L_n(\varphi; A, X) \qquad (4.4.12)$$

i.e. the quadrature formula (4.4.6) is the best among all quadrature formulas of Niederreiter type (4.4.2)–(4.4.4).

### Proof

Let $n$ and $r$ be positive integers. Then for each function $f \in W^r H^\omega$

$$|R_r(f; A, X)| = \left| \int_0^1 f(x)\mathrm{d}x - Q^r(f; A, X) \right|$$

$$\overset{\substack{(4.4.3)\\(4.4.4)}}{=} \left| \sum_{i=1}^n \int_{a_{i-1}}^{a_i} \left[ f(x) - \sum_{k=0}^r \frac{f^{(k)}(x_i)}{k!}(x - x_i)^k \right] \mathrm{d}x \right|$$

$$\overset{(2.5.1)}{=} \left| \sum_{i=1}^n \int_{a_{i-1}}^{a_i} \frac{(x - x_i)^r}{(r-1)!} \int_0^1 (1 - z)^{r-1} \right.$$

$$\left. \times [f^{(r)}(x_i + z(x - x_i)) - f^{(r)}(x_i)]\mathrm{d}z \, \mathrm{d}x \right|$$

$$\leqq \sum_{i=1}^n \int_{a_{i-1}}^{a_i} \frac{|x - x_i|^r}{(r-1)!} \int_0^1 (1 - z)^{r-1} \cdot \omega(z|x - x_i|)\mathrm{d}z \, \mathrm{d}x$$

$$
\overset{(4.4.9)}{=} \sum_{i=1}^{n} \int_{a_{i-1}}^{a_i} \varphi(|x - x_i|)\mathrm{d}x \overset{(2.1.2)}{=} L_n(\varphi; A, X). \tag{4.4.13}
$$

Therefore

$$
\sup\{|R_r(f; A, X)| : f \in W^r H^\omega\} \leqq L_n(\varphi; A, X). \tag{4.4.14}
$$

In particular, it holds that

$$
\sup\{|R_r(f; B, X)| : f \in W^r H^\omega\} \leqq L_n(\varphi; B, X). \tag{4.4.15}
$$

On the other hand, by Lemma 2.5.1, the function $f_r(x) = f_r(\omega, X; x)$, defined in (2.5.2), belongs to the function class $W^r H^\omega$ and for *even r* equation (2.5.3) holds. From (2.5.3) and (4.4.7) it follows that

$$
R_r(f_r; B, X) = \int_0^1 f_r(x)\mathrm{d}x - Q^r(f_r; B, X)
$$

$$
\overset{(4.4.7)}{=} \int_0^1 [f_r(x) - \Phi^r(f_r, X; x)]\mathrm{d}x
$$

$$
\overset{(2.5.3)}{=} \sum_{i=1}^{n} \int_{b_{i-1}}^{b_i} \frac{|x - x_i|^r}{(r-1)!} \int_0^1 (1-z)^{r-1}\omega(z|x - x_i|)\mathrm{d}x
$$

$$
\overset{(4.4.9)}{=} \sum_{i=1}^{n} \int_{b_{i-1}}^{b_i} \varphi(|x - x_i|)\mathrm{d}x = L_n(\varphi; B, X). \tag{4.4.16}
$$

From (4.4.15) and (4.4.16) we get

$$
\sup\{|R_r(f; B, X)| : f \in W^r H^\omega\} = R_r(f_r; B, X) = L_n(\varphi; B, X)
$$

and thus (4.4.8) is proved.

Now let equation (4.4.10) hold. From it and (4.4.8) it is clear that inequality (4.4.11) can be proved provided we first prove that

$$
L_n(\varphi; A, X) \geqq L_n(\varphi; B, X). \tag{4.4.17}
$$

If $A = B$, then (4.4.17) is satisfied. Let $a_i \neq b_i$ $(1 \leq i \leq n - 1)$, and all the remaining nodes of mesh $A$ coincide with the corresponding nodes of mesh $B$. (The general case reduces to this case.) Then, $L_n(\varphi; A, X)$ and $L_n(\varphi; B, X)$ are different only in the expressions

$$\alpha = \int_{x_i}^{a_i} \varphi(|x - x_i|)dx + \int_{a_i}^{x_{i+1}} \varphi(|x - x_{i+1}|)dx$$

and

$$\beta = \int_{x_i}^{b_i} \varphi(|x - x_i|)dx + \int_{b_i}^{x_{i+1}} \varphi(|x - x_{i+1}|)dx.$$

Let us consider the case $a_i < b_i$. (The case $a_i > b_i$ can be treated in the same way.)

From (4.4.9) it follows that the function $\varphi(u)$ is strictly increasing in the segment $[0, 1]$. Therefore,

$$L_n(\varphi; A, X) - L_n(\varphi; B, X) = \alpha - \beta = \int_{a_i}^{b_i} [\varphi(|x - x_{i+1}|) - \varphi(|x - x_i|)]dx > 0$$

since for each $x \in [a_i, b_i]$ the inequality $|x - x_{i+1}| \geq |x - x_i|$ holds with equality only for $x = b_i$. Hence $\varphi(|x - x_{i+1}|) \geq \varphi(|x - x_i|)$ with equality for $x = b_i$ only.

Thus (4.4.11) is proved. Since $(B, X) \in (A, X)_0$, then (4.4.11) implies (4.4.12). The lemma is proved.


**Corollary 4.4.1** *Let $n \in \mathbb{N}$; $r(r \in \mathbb{N})$ be an even number; $X \in (X)_0$ and $\omega(t)$ $(0 \leq t \leq 1)$ be an arbitrary modulus of continuity. Then*

$$\Delta(W^rH^\omega, X) = \Delta(W^rH^\omega, X)_{L_1} \tag{4.4.18}$$

*holds, where $\Delta(W^rH^\omega, X)_{L_1}$ is determined by (1.5.28) for $p = 1$.*


Indeed, equation (4.4.18) follows from (2.5.7) (for $p = 1$) and from (4.4.8).


**Theorem 4.4.1**


*Let $n \in \mathbb{N}$, $r(r \in \mathbb{N})$ be an even number, $v$, be a fixed element of the set $\{0, 1, 2, 3\}$, $X \in (X)_v$, and $\omega(t)$ $(0 \leq t \leq 1)$ be an arbitrary modulus of continuity.*

*Out of all the quadrature formulas of Niederreiter's type (4.4.6), using only the information $T_r(f, X)$, $X \in (X)_v$, the optimal one in the class $W^r H^\omega$ $(r = 2, 4, 6, \ldots)$ is the formula*

$$\int_0^1 f(x)\mathrm{d}x = Q^r(f; A_v, X_v) + R_r(f; A_v, X_v) \tag{4.4.19}$$

*constructed according to formulas (4.4.2)–(4.4.4) by means of the mesh pair†*
*$(A_v, X_v)$, defined in $(1.6.5_v)$.*

    *Furthermore,*

$$\Delta_n^v(W^r H^\omega) \overset{\text{def}}{=} \inf_{X \in (X)_v} \Delta(W^r H^\omega, X)$$

$$= \Delta(W^r H^\omega, X_v)$$

$$= \sup\left\{ \left| \int_0^1 f(x)\mathrm{d}x - Q^r(f; A_v, X_v) \right| : f \in W^r H^\omega \right\}$$

$$= \int_0^1 f_r(x)\mathrm{d}x - Q^r(f_r; A_v, X_v)$$

$$= \frac{D^r(A_v, X_v)_\infty}{(\tau - 1)!} \cdot \int_0^1\int_0^1 u^r(1 - z)^{r-1} \cdot \omega(zuD(A_v, X_v)_\infty)\mathrm{d}z\,\mathrm{d}u \tag{4.4.20}$$

*where $D(A, X)_\infty$ is the uniform deviation of the mesh pair $(A, X)$.*
*The function $f_r(\omega, X_v; x)$, defined by (2.5.2) for $X = X_v$ is extremal.*

### Proof

By the definition

$$\Delta_n^v(W^r H^\omega) = \inf\{\Delta(W^r H^\omega; X) : X \in (X)_v\}$$

from Corollary 4.4.1 we have

$$\inf\{\Delta(W^r H^\omega; X) : X \in (X)_v\} = \inf\{\Delta(W^r H^\omega; X)_{L_1} : X \in (X)_v\}$$

† It can be easily seen that $A_v$ is uniquely determined by $X_v$ using formulas (4.4.5).

and from Theorem 2.5.2 it follows that

$$\inf\{\Delta(W^rH^\omega; X)_{L_1}: X \in (X)_v\} = \Delta(W^rH^\omega; X_v)_{L_1}.$$

This relation implies

$$\Delta_n^v(W^rH^\omega) = \Delta(W^rH^\omega; X_v)_{L_1} = \Delta(W^rH^\omega; X_v) \qquad (4.4.21)$$

again using Corollary 4.4.1.

Now, from (4.4.21), (4.4.8) and (2.5.16) (for $p = 1$) it follows that

$$\Delta_n^v(W^rH^\omega) = \Delta(W^rH^\omega; X_v) = \sup_{f \in W^rH^\omega} |R_r(f; A_v, X_v)|$$

$$= R_r(f_r; A_v, X_v) = \Delta(W^rH^\omega; X_v)_{L_1}$$

$$= \frac{D^r(A_v, X_v)_\infty}{(r-1)!} \int_0^1\!\!\int_0^1 u^r(1-z)^{r-1}\omega(zuD(A_v, X_v)_\infty)\,\mathrm{d}z\,\mathrm{d}u.$$

Thus, equalities (4.4.20) are proved.

Equations (4.4.20) imply all the assertions of the theorem, since by definition (4.4.21) means that the mesh $X_v$ is optimal, and that quadrature formula (4.4.19) is optimal in the class $W^rH^\omega$ ($\tau = 2, 4, 6, \ldots$) among all the quadrature formulas of Niederreiter type (4.4.6); in (4.4.21) information for the extremality of the function $f_r(\omega, X_v; x)$ is contained, and for the error in the class $W^rH^\omega$ also.

The theorem is proved.

We will state the assertions of the theorem in the cases $v = 1$, $v = 2$ and $v = 3$ in the following corollaries.

**Corollary 4.4.2**  *Let $n \in \mathbb{N}$, $r = 2, 4, 6, \ldots$, and $\omega(t)$ $(0 \leq t \leq 1)$ be an arbitrary modulus of continuity. Then, from all quadrature formulas of form (4.4.6), using only the information $T_r(f, X)$, $X \in (X)_0$ the optimal in the class $W^rH^\omega$ $(r = 2, 4, 6, \ldots)$ is the formula†*

$$\int_0^1 f(x)\mathrm{d}x = Q^r(f; A_0, X_0) + R_r(f; A_0, X_0) \qquad (4.4.22)$$

† The integer part of $\alpha$ is denoted by $[\alpha]$.

$$Q^r(f; A_0, X_0) = \frac{1}{n} \sum_{i=1}^{n} \sum_{k=0}^{[r/2]} \frac{f^{(2k)}((2i-1)/2n)}{(2n)^{2k}(2k+1)!} \qquad (4.4.23)$$

*with the error*

$$\Delta_n^0(W^r H^\omega) = R_r(f_r; A_0, X_0)$$

$$= \frac{1}{(2n)^r(r-1)!} \int_0^1 \int_0^1 u^r(1-z)^{r-1}\omega(uz/2n)\,dz\,du \qquad (4.4.24)$$

*where the mesh pair* $(A_0, X_0)$ *is determined in* $(1.6.5_0)$. *The function* $f_r(\omega, X_0; x)$, *defined by* $(2.5.2)$ *for* $X = X_0$ *is extremal.*

### Proof

Since, according to $(2.1.17_0)$ $D(A_0, X_0)_\infty = 1/2n$, then the assertions of the corollary are implied by the assertions of the theorem. It is sufficient only to calculate coefficients in $(4.4.23)$. From $(4.4.4)$ and $(1.6.50)$ it follows $(v = 0)$ that

$$\lambda_{ik}^{(0)} = \frac{(a_i^{(0)} - x_i^{(0)})^{k+1} - (a_{i-1}^{(0)} - x_i^{(0)})^{k+1}}{(k+1)!}$$

$$= \frac{\left(\dfrac{i}{n} - \dfrac{2i-1}{2n}\right)^{k+1} - \left(\dfrac{i-1}{n} - \dfrac{2i-1}{2n}\right)^{k+1}}{(k+1)!} = \frac{1-(-1)^{k+1}}{(2n)^{k+1}(k+1)!}$$

or

$$\lambda_{ik}^{(0)} = \begin{cases} \dfrac{1}{n(2n)^k(k+1)!} & \text{for even } k, \\[2ex] 0 & \text{for odd } k. \end{cases}$$

**Corollary 4.4.3** *Let* $n \in \mathbb{N}$, $r = 2, 4, 6, \ldots$, *and* $\omega(t)$ $(0 \leq t \leq 1)$ *be an arbitrary modulus of continuity. Then, from all quadrature formulas of the form* $(4.4.6)$ *using only information* $T_r(f, X)$, $X \in (X)_1$, *the optimal formula in the class* $W^r H^\omega$ *is*

$$\int_0^1 f(x)\,dx = Q^r(f; A_1, X_1) + R_r(f; A_1, X_1) \qquad (4.4.25)$$

$$Q^r(f; A_1, X_1) = \frac{1}{2n-1} \sum_{k=0}^{r} \frac{f^{(k)}(0)}{(2n-1)^k(k+1)!}$$

$$+ \frac{2}{2n-1} \sum_{i=2}^{n} \sum_{k=0}^{[r/2]} \frac{f^{(2k)}\left(\frac{2i-2}{2n-1}\right)}{(2n-1)^{2k}(2k+1)!} \qquad (4.4.26)$$

*with the error*

$$\Delta_n^1(W^r H^\omega) = \inf_{X \in (X)_1} \sup_{f \in W^r H^\omega} |R_r(f; B, X)| = R_r(f_r; A_1, X_1)$$

$$= \frac{1}{(2n-)^r(r-1)!} \int_0^1 \int_0^1 u^r(1-z)^{r-1}\omega\left(\frac{zu}{2n-1}\right) dz\, du \qquad (4.4.27)$$

*where the mesh pair $(A_1, X_1)$ is determined by $(1.6.5_1)$. The function $f_r(\omega, X_1; x)$, defined by $(2.5.2)$ for $X = X_1$ is extremal.*

Indeed, by $(2.1.17_1)$

$$D(A_1, X_1)_\infty = 1/(2n-1)$$

and by $(1.6.5_1)$ and $(4.4.4)$ it follows that

$$\lambda_{1k}^{(1)} = \frac{(a_1^{(1)} - x_1^{(1)})^{k+1} - (a_1^{(1)} - x_1^{(1)})^{k+1}}{(k+1)!}$$

$$= \frac{1}{(2n-1)^{k+1}(k+1)!} \qquad (k=0, 1, \ldots, r)$$

$$\lambda_{ik}^{(1)} = \frac{(a_i^{(1)} - x_i^{(1)})^{k+1} - (a_{i-1}^{(1)} - x_i^{(1)})^{k+1}}{(k+1)!}$$

$$= \begin{cases} \dfrac{2}{(2n-1)(2n-1)^k(k+1)!} & \text{for even } k, \\ 0 & \text{for odd } k. \end{cases}$$

$(i = 2, 3, \ldots, n; k = 0, 1, \ldots, r)$.

The next part follows from Theorem 4.4.1.

**Corollary 4.4.4**   *Let $n = 2, 3, 4, \ldots, r = 2, 4, 6, \ldots,$ and $\omega(t)$ $(0 \leq t \leq 1)$ be an arbitrary modulus of continuity. From all quadrature formulas of the form (4.4.6), using only the information $T_r(f; X)$, $X \in (X)_2$, optimal formula in the class $W^r H^\omega$ is*

$$\int\limits_0^1 f(x)\mathrm{d}x = Q^r(f; A_2, X_2) + R_r(f; A_2, X_2) \tag{4.4.28}$$

$$Q^r(f; A_2, X_2) = \frac{1}{2n-2} \sum_{k=0}^{r} \frac{f^{(k)}(0) + (-1)^k f^{(k)}(1)}{(2n-2)^k (k+1)!}$$

$$+ \frac{1}{n-1} \sum_{i=2}^{n-1} \sum_{k=0}^{[r/2]} \frac{f^{(2k)}\left(\dfrac{i-1}{n-1}\right)}{(2n-2)^{2k}(2k+1)!} \tag{4.4.29}$$

*with the error*

$$\Delta_n^2(W^r H^\omega) = \inf_{X \in (X)_2} \ \sup_{f \in W^r H^\omega} |R_r(f; B, X)|$$

$$= R_r(f_r; A_2, X_2)$$

$$= \frac{1}{(2n-2)^r(r-1)!} \int\limits_0^1\int\limits_0^1 u^r(1-z)^{r-1}\omega\left(\frac{uz}{2n-2}\right)\mathrm{d}z\,\mathrm{d}u \tag{4.4.30}$$

*where the mesh pair $(A_2, X_2)$ is determined by $(1.6.5_2)$. The function $f_r(\omega, X_2; x)$, defined by $(2.5.2)$ for $X = X_2$ is extremal.*

Indeed, from $(2.1.17_2)$ it follows that

$$d(A_2, X_2)_\infty = 1/(2n-2)$$

and from $(1.6.5_2)$ and $(4.4.4)$ it follows correspondingly that

$$\lambda_{1k}^{(2)} = \frac{(a_1^{(2)} - x_1^{(2)})^{k+1} - (a_0^{(2)} - x_1^{(2)})^{k+1}}{(k+1)!}$$

$$= \frac{1}{(2n-2)^{k+1}(k+1)!}, \qquad k = 0, 1, \ldots, r;$$

$$\lambda_{nk}^{(2)} = \frac{(a_n^{(2)} - x_n^{(2)})^{k+1} - (a_{n-1}^{(2)} - x_n^{(2)})^{k+1}}{(k+1)!}$$

$$= \frac{(-1)^k}{(2n-2)^{k+1}(k+1)!}, \qquad k = 0, 1, \ldots, r;$$

$$\lambda_{ik}^{(2)} = \frac{(a_i^{(2)} - x_i^{(2)})^{k+1} - (a_{i-1}^{(2)} - x_i^{(2)})^{k+1}}{(k+1)!}$$

$$= \begin{cases} \dfrac{1}{(n-1)(2n-2)^k(k+1)!} & \text{for even } k, \\[2mm] 0 & \text{for odd } k. \end{cases}$$

$(i = 2, 3, \ldots, n-1; \; k = 0, 1, \ldots, r)$.

The next part follows from Theorem 4.4.1.

**Corollary 4.4.5** *Let $n \in \mathbb{N}$, $r = 2, 4, 6, \ldots$, and $\omega(t)$ $(0 \leq t \leq 1)$ be an arbitrary modulus of continuity.*

*From all quadrature formulas of the form (4.4.6), using only, the information $T_r(f, X)$, $X \in (X)_3$, in the class $W^r H^\omega$ the optimal formula is*

$$\int_0^1 f(x)\mathrm{d}x = Q^r(f; A_3, X_3) + R_2(f; A_3, X_3), \tag{4.4.31}$$

$$Q^r(f; A_3, X_3) = \frac{2}{2n-1} \sum_{i=1}^{n-1} \sum_{k=0}^{[r/2]} \frac{f^{(2k)}\left(\dfrac{2i-1}{2n-1}\right)}{(2n-1)^{2k}(2k+1)!}$$

$$+ \frac{1}{2n-1} \sum_{k=0}^{r} \frac{(-1)^k f^{(k)}(1)}{(2n-1)^k(k+1)!} \tag{4.4.32}$$

*with the error*

$$\Delta_n^3(W^r H^\omega) = \inf_{X \in (X)_3} \sup_{f \in W^r H^\omega} |R_r(f; B, X)|$$

$$= R_r(f_2; A_3, X_3)$$

$$= \frac{1}{(2n-1)^r(r-1)!} \int_0^1 \int_0^1 u^r (1-z)^{r-1} \omega\left(\frac{uz}{2n-1}\right) \mathrm{d}z \, \mathrm{d}u \tag{4.4.33}$$

*where the mesh pair $(A_3, X_3)$ is determined by $(1.6.5_3)$. The function $f_r(\omega, X_3; x)$, defined by $(2.5.1)$ for $X = X_3$ is extremal.*

Indeed, from $(2.1.17_3)$ if follows that

$$D(A_3, X_3)_\infty = 1/(2n - 1)$$

and from $(1.6.5_3)$ and $(4.4.4)$ it follows correspondingly that

$$\lambda_{nk}^{(3)} = \frac{(a_n^{(3)} - x_n^{(3)})^{k+1} - (a_{n-1}^{(3)} - x_n^{(3)})^{k+1}}{(k+1)!}$$

$$= \frac{(-1)^k}{(2n-1)^{k+1}(k+1)!}, \qquad k = 0, 1, \ldots, r;$$

$$\lambda_{ik}^{(3)} = \frac{(a_i^{(3)} - x_i^{(3)k+1} - (a_{i-1}^{(3)} - x_i^{(3)})^{k+1}}{(k+1)!}$$

$$= \begin{cases} \dfrac{2}{(2n-1)^{k+1}(k+1)!} & \text{for even } k, \\[2mm] 0 & \text{for odd } k. \end{cases}$$

$(i = 1, 2, \ldots, n - 1;\ k = 0, 1, \ldots, r)$.

For a quadrature formula

$$\int_0^1 f(x)\mathrm{d}x = Q_n(f) + R_n(f)$$

it is said [27] that it has an algebraic degree of precision $m$, if for every polynomial $P(x)$ of degree $\leq m$ the equality $R_n(P) = 0$ holds and if there exists at least one polynomial $P_n(x)$ of degree $m + 1$ for which $R_n(P_1) \neq 0$.

Finally, we will consider the question for the algebraic degree of precision of quadrature formula (4.4.22), (4.4.23).

**Proposition 4.4.1**     *Let $n \in \mathbb{N}$ and $r = 2, 4, 6, \ldots$ The algebraic degree of precision of formula (4.4.22), (4.4.23) is equal to $r + 1$.*

The proof of this assertion will be accomplished, if we show that for each $k = 0, 1, \ldots, r, r + 1$ the equality

$$R_r(x^k; A_0, X_0) = 0 \qquad (4.4.34)$$

is satisfied, and (for $k = r + 2$) the inequality

$$R_r(x^{r+2}; A_0, X_0) \neq 0. \qquad (4.4.35)$$

holds.

According to (4.4.13) and ($1.6.5_0$) we have

$$R_r(f; A_0, X_0)$$

$$= \int_0^1 f(x)dx - Q^r(f; A_0, X_0)$$

$$= \sum_{i=1}^n \int_{a_{i-1}^{(0)}}^{a_i^{(0)}} \frac{(x - x_i^{(0)})^r}{(r-1)!} \int_0^1 (1-z)^{r-1}[f^{(r)}(x_i^{(0)} + z(x - x_i^{(0)})) - f^{(r)}(x_i^{(0)})]dz\, dx$$

$$= \sum_{i=1}^n \int_{-1/2n}^{1/2n} \frac{u^r}{(r-1)!} \int_0^1 (1-z)^{r-1}\left[ f^{(r)}\left(\frac{2i-1}{2n} + zu\right) - f^{(r)}\left(\frac{2i-1}{2n}\right)\right]dz\, du.$$

$$(4.4.36)$$

(a) Let $f(x) = x^k$ ($k = 0, 1, 2, \ldots, r$). Then, for each $i = 1, 2, \ldots, n$

$$f^{(r)}\left(\frac{2i-1}{2n} + zu\right) - f^{(r)}\left(\frac{2i-1}{2n}\right) = 0$$

and hence from (4.4.36) follows (4.4.34) for each $k = 0, 1, 2, \ldots, r$.

(b) Let $f(x) = x^{r+1}$. Then $f^{(r)}(x) = (r+1)!x$ and for each $i = 1, 2, \ldots, n$

$$f^{(r)}\left(\frac{2i-1}{2n} + zu\right) - f^{(r)}\left(\frac{2i-1}{2n}\right) = (r+1)!zu.$$

These relations and (4.4.36) imply

$$R_r(x^{r+1}; A_0, X_0) = nr(r+1) \int_{-1/2n}^{1/2n} u^{r+1}\, du \int_0^1 z(1-z)^{r-1}\, dz = 0$$

since, by hypothesis $r$ is an even number and hence (for odd $r+1$)

$$\int_{-1/2n}^{1/2n} u^{r+1}\, du = 0. \tag{4.4.37}$$

Thus (4.4.34) is proved for $k = r + 1$.

(c) Let $f(x) = x^{r+2}$. Then

$$f^{(r)}\left(\frac{2i-1}{2n} + zu\right) - f^{(r)}\left(\frac{2i-1}{2n}\right) = \frac{(r+2)!}{2}\left(z^2u^2 + \frac{2i-1}{n}\,zu\right).$$

This relation along with (4.4.36) imply

$$R_r(x^{r+2}; A_0, X_0)$$

$$= n \cdot \frac{r(r+1)(r+2)}{2} \int_{-1/2n}^{1/2n} u^{r+2}\, du \int_0^1 z^2(1-z)^{r-1}\, dz$$

$$= \frac{1}{(r+3)(2n)^{r+2}} \neq 0$$

again using (4.4.37). Thus the assertion is proved.

**5**

---

# Optimal cubature formulas with restrictions on the lattice for the function classes $H^{\omega_1\omega_2}(D_2)$ and $H^{\omega}(D_2)$

In this chapter we give a proof of the statements, announced in [30] along with proofs of propositions of the same sort, not included in [30]. We use the notions, notation and results of Chapter 3.

## 5.1  Best cubature formulae

### Lemma 5.1.1

*Let $n$ and $m$ be fixed positive integers, $X = (x_1, x_2, \ldots, x_n)$ and $Y = (y_1, y_2, \ldots, y_m)$ be arbitrary meshes of nodes in the segment $[0, 1]$, $(X, Y) \in (X, Y)_{0,0}$, and $\omega_1(t)$, $\omega_2(t)$ and $\omega(t)$ be arbitrary moduli of continuity. Then*

$$\sup\left\{\int_0^1\int_0^1 f(x, y)\mathrm{d}x\,\mathrm{d}y : f \in H^{\omega_1\omega_2}(D_2)_0\right\} = \int_0^1\int_0^1 f_{12}(X, Y; x, y)\mathrm{d}x\,\mathrm{d}y \quad (5.1.1)$$

*and*

$$\sup\left\{\int_0^1\int_0^1 f(x, y)\mathrm{d}x\,\mathrm{d}y : f \in H^{\omega}(D_2)_0\right\} = \int_0^1\int_0^1 f_0(X, Y; x, y)\mathrm{d}x\,\mathrm{d}y \quad (5.1.2)$$

*where $f_{12}(X, Y; x, y)$ and $f_0(X, Y; x, y)$ are the standard functions, defined in (3.1.7) and (3.1.8) respectively, and the function classes $H^{\omega_1\omega_2}(D_2)_0$ and $H^{\omega}(D_2)_0$ in (3.1.4) and (3.1.5) respectively.*

## Proof

According to Lemma 3.1.2 (see (3.1.9)) for every function $f \in H^{\omega_1\omega_2}(D_2)_0$ and for each point $(x, y) \in D_2$ the inequality

$$f(x, y) \leqq f_{12}(X, Y; x, y)$$

is satisfied. Integrating this inequality on $D_2$, we get

$$\int_0^1\int_0^1 f(x, y)\,dx\,dy \leqq \int_0^1\int_0^1 f_{12}(X, Y; x, y)\,dx\,dy.$$

Since, again by Lemma 3.1.2, $f_{12} \in H^{\omega_1\omega_2}(D_2)_0$, then the above inequality implies that

$$\sup\left\{\int_0^1\int_0^1 f(x, y)\,dx\,dy : f \in H^{\omega_1\omega_2}(D_2)_0\right\} = \int_0^1\int_0^1 f_{12}(X, Y; x, y)\,dx\,dy$$

and thus (5.1.1) is proved.

In the same manner, again using Lemma 3.1.2, we can prove equality (5.1.2).

The lemma is proved.

**Corollary 5.1.1**    *Under the hypothesis and notation of Lemma 5.1.1, it follows that*

$$\sup\left\{\int_0^1\int_0^1 f(x, y)\,dx\,dy : f \in H^{\omega_1\omega_2}(D_2)_0\right\} \leqq \omega_1(1) + \omega_2(1) < +\infty \quad (5.1.3)$$

*and*

$$\sup\left\{\int_0^1\int_0^1 f(x, y)\,dx\,dy : f \in H^{\omega}(D_2)_0\right\} \leqq \omega(\sqrt{2}) < +\infty \qquad (5.1.4)$$

Indeed, according to (5.1), (3.1.7) and the monotonicity of the moduli of continuity, we get

$$\sup\left\{\int_0^1\int_0^1 f(x,y)\mathrm{d}x\,\mathrm{d}y:\ f\in H^{\omega_1\omega_2}(D_2)_0\right\}$$

$$=\int_0^1\int_0^1 f_{12}(X,Y;x,y)\mathrm{d}x\,\mathrm{d}y$$

$$=\sum_{i=1}^n\int_{b_{i-1}}^{b_i}\omega_1(|x-x_i|)\mathrm{d}x+\sum_{j=1}^m\int_{e_{j-1}}^{e_j}\omega_2(|y-y_j|)\mathrm{d}y$$

$$\leqq\omega_1(1)\sum_{i=1}^n(b_i-b_{i-1})+\omega_2(1)\sum_{j=1}^m(e_j-e_{j-1})$$

$$=\omega_1(1)+\omega_2(1)<+\infty. \tag{5.1.4'}$$

Thus (5.1.3) is proved. In a similar manner one can prove (5.1.4) also. The proof is completed.

## Lemma 5.1.2

*Let $n\in\mathbb{N}$, $m\in\mathbb{N}$, $(X,Y)\in(X,Y)_{0,0}$ be an aribtrary lattice of points in the unit square $D_2\subset\mathbb{R}^2$, and $\omega_1(t)$, $\omega_2(t)$ $(0\leq t\leq 1)$ and $\omega(t)$ be arbitrary moduli of continuity. Then for the errors of the best cubature formulas for approximate calculation of the integral*

$$Lf=\int_0^1\int_0^1 f(x,y)\mathrm{d}x\,\mathrm{d}y$$

*in the classes $H^{\omega_1\omega_2}(D_2)$ and $H^\omega(D_2)$ on information $T(f;X,Y)$, $(X,Y)\in(X,Y)_{0,0}$ are given by the formulas*

$$R(H^{\omega_1\omega_2}(D_2);X,Y)=\int_0^1\int_0^1 f_{12}(X,Y;x,y)\mathrm{d}x\,\mathrm{d}y \tag{5.1.5}$$

*and*

$$R(H^\omega(D_2); X, Y) = \int_0^1\int_0^1 f_0(X, Y; x, y)\,dx\,dy \qquad (5.1.6)$$

*where $f_{12}(X, Y; x, y)$ and $f_0(X, Y; x, y)$ are the standard functions (3.1.7) and (3.1.8) respectively.*

## Proof

From the results of Section 3.1 and Corollary 5.1.1 it follows that all the hypotheses of Smolyak's lemma for the linear functionals

$$Lf = \int_0^1\int_0^1 f(x, y)\,dx\,dy$$

and

$$L_{ij}f = f(x_i, y_j) \quad (i = 1, 2, \ldots, n; j = 1, 2, \ldots, m)$$

and the function classes $H^{\omega_1\omega_2}(D_2)$ and $H^\omega(D_2)$ are satisfied in the normed linear space $C(D_2)$. According to a corollary of this lemma, the following equations are satisfied

$$\sup\left\{\int_0^1\int_0^1 f(x, y)\,dx\,dy : f \in H^{\omega_1\omega_2}(D_2)_0\right\} = R(H^{\omega_1\omega_2}(D_2); X, Y) \quad (5.1.7)$$

and

$$\sup\left\{\int_0^1\int_0^1 f(x, y)\,dx\,dy : f \in H^\omega(D_2)_0\right\} = R(H^\omega(D_2); X, Y) \qquad (5.1.8)$$

where $R(H^{\omega_1\omega_2}(D_2); X, Y)$ and $R(H^\omega(D_2); X, Y)$ are the corresponding errors of the best cubature formulas for $Lf$ in $H^{\omega_1\omega_2}(D_2)$ and $H^\omega(D_2)$ on information $T(f; X, Y)$, $(X, Y) \in (X, Y)_{0,0}$. Then (5.1.1) and (5.1.7) imply (5.1.5), and (5.1.2) and (5.1.8) imply (5.1.6), respectively.

The lemma is proved.

Let $\Phi(f; X, Y; x, y)$ be the best method (3.1.15) for recovery of functions of the classes $H^{\omega_1\omega_2}(D_2)$ and $H^{\omega}(D_2)$ on information $T(f; X, Y), (X, Y)\in(X, Y)_{0,0}$.

Further, we are interested in the cubature formula

$$\int_0^1\int_0^1 f(x, y)\mathrm{d}x\,\mathrm{d}y = Q(f; X, Y) + R(f; x, Y) \qquad (5.1.9)$$

where

$$Q(f; X, Y) = \int_0^1\int_0^1 \Phi(f; X, Y; x, y)\mathrm{d}x\,\mathrm{d}y$$

$$= \sum_{i=1}^{n}\sum_{j=1}^{m} (b_i - b_{i-1})(e_j - e_{j-1})f(x_i, y_j). \qquad (5.1.10)$$

### Theorem 5.1.1

*Let $n\in\mathbb{N}$, $m\in\mathbb{N}$, $(X, Y)\in(X, Y)_{0,0}$ and $\omega_1(t)$, $\omega_2(t)$ and $\omega(t)$ be arbitrary moduli of continuity. Then from all possible cubature formulas for approximate calculation of the integral*

$$\int_0^1\int_0^1 f(x, y)\mathrm{d}x\,\mathrm{d}y$$

*using only the information $T(f; X, Y)$, $(X, Y)\in(X, Y)_{0,0}$, the best formula in the function classes $H^{\omega_1\omega_2}(D_2)$ and $H^{\omega}(D_2)$ is (5.1.9), (5.1.10) with the corresponding errors*

$$R(H^{\omega_1\omega_2}(D_2); X, Y) = \sup\left\{|R(f; X, Y)|: f\in H^{\omega_1\omega_2}(D_2)\right\}$$

$$= R(f_{12}; X, Y)$$

$$= \int_0^1\int_0^1 f_{12}(X, Y; x, y)\mathrm{d}x\,\mathrm{d}y \qquad (5.1.11)$$

*and*

$$R(H^{\omega}(D_2); X, Y) = \sup\{|R(f; X, Y)|: f \in H^{\omega}(D_2)\}$$

$$= R(f_0; X, Y)$$

$$= \int_0^1 \int_0^1 f_0(X, Y; x, y)\mathrm{d}x\,\mathrm{d}y \qquad (5.1.12)$$

*where $f_{12}(X, Y; x, y)$ and $f_0(X, Y; x, y)$ are the standard functions (3.1.7) and (3.1.8) respectively.*

## Proof

For every function $f \in H^{\omega_1 \omega_2}(D_2)$ we obtain from (5.1.9), (5.1.10) and (3.1.15) the following estimate

$$|R(f; X, Y)| \overset{(5.1.9)}{=} \left| \int_0^1 \int_0^1 f(x, y)\mathrm{d}x\,\mathrm{d}y - Q(f; X, Y) \right|$$

$$\overset{(5.1.0)}{=} \left| \int_0^1 \int_0^1 f(x, y)\mathrm{d}x\,\mathrm{d}y - \int_0^1 \int_0^1 \Phi(f; X, Y; x, y)\mathrm{d}x\,\mathrm{d}y \right|$$

$$\overset{(3.1.15)}{=} \left| \sum_{i=1}^{n} \sum_{j=1}^{m} \int_{b_{i-1}}^{b_i} \int_{e_{j-1}}^{e_j} [f(x, y) - f(x_i, y_j)]\mathrm{d}x\,\mathrm{d}y \right|$$

$$\leq \sum_{i=1}^{n} \sum_{j=1}^{m} \int_{b_{i-1}}^{b_i} \int_{e_{j-1}}^{e_j} [\omega_1(|x - x_i|) + \omega_2(|y - y_j|)]\mathrm{d}x\,\mathrm{d}y$$

$$= \int_0^1 \int_0^1 \sum_{i=1}^{n} \sum_{j=1}^{m} \chi_i(x)\chi_j(y)[\omega_1(|x - x_i|) + \omega_2(|y - y_j|)]\mathrm{d}x\,\mathrm{d}y$$

$$\overset{(3.1.7)}{=} \int_0^1 \int_0^1 f_{12}(X, Y; x, y)\mathrm{d}x\,\mathrm{d}y.$$

Therefore

$$\sup\{|R(f;X,Y)|: f \in H^{\omega_1\omega_2}(D_2)\} \leq \int_0^1\int_0^1 f_{12}(X,Y;x,y)\mathrm{d}x\,\mathrm{d}y. \quad (5.1.13)$$

According to Lemma 3.1.2 $f_{12}(X,Y;x,y) \in H^{\omega_1\omega_2}(D_2)_0$ and hence $Q(f_{12};X,Y) = 0$ and

$$\int_0^1\int_0^1 f_{12}(X,Y;x,y)\mathrm{d}x\,\mathrm{d}y = R(f_{12};X,Y).$$

Taking into account that $f_{12} \in H^{\omega_1\omega_2}(D_2)$, from this relation and from (5.1.13) we obtain the equations

$$\sup\{|R(f;X,Y)|: f \in H^{\omega_1\omega_2}(D_2)\} = R(f_{12};X,Y) = \int_0^1\int_0^1 f_{12}(X,Y;x,y)\mathrm{d}x\,\mathrm{d}y.$$

$$(5.1.14)$$

Then, from (5.1.14) and (5.1.5) we obtain (5.1.11).

Relations (5.1.12) can be proved in an analogous manner. Indeed, for each function $f \in H^{\omega}(D_2)$ from (5.1.9), (5.1.10) and (3.1.15) the estimate

$$|R(f;X,Y)| = \left|\int_0^1\int_0^1 f(x,y)\mathrm{d}x\,\mathrm{d}y - Q(f;X,Y)\right|$$

$$= \left|\sum_{i=1}^{n}\sum_{j=1}^{m}\int_{b_{i-1}}^{b_i}\int_{e_{j-1}}^{e_j}[f(x,y) - f(x_i,y_j)]\mathrm{d}x\,\mathrm{d}y\right|$$

$$\leq \sum_{i=1}^{n}\sum_{j=1}^{m}\int_{b_{i-1}}^{b_i}\int_{e_{j-1}}^{e_j}\omega(\rho(M,M_{ij}))\mathrm{d}x\,\mathrm{d}y$$

$$= \int_0^1\int_0^1 \sum_{i=1}^{n}\sum_{j=1}^{m}\chi_i(x)\chi_j(y)\omega(\rho(M,M_{ij}))\mathrm{d}x\,\mathrm{d}y$$

$$(3.1.8) \quad = \int_0^1 \int_0^1 f_0(X, Y; x, y)\,dx\,dy$$

follows.

Therefore

$$\sup\{|R(f; X, Y)| : f \in H^\omega(D_2)\} \leq \int_0^1 \int_0^1 f_0(X, Y; x, y)\,dx\,dy. \quad (5.1.15)$$

According to Lemma 3.1.2, $f_0(X, Y; x, y) \in H^\omega(D_2)_0$ and hence $Q(f_0; X, Y) = 0$ and

$$\int_0^1 \int_0^1 f_0(X, Y; x, y)\,dx\,dy = R(f_0; X, Y).$$

From this relation and from (5.1.15), taking into account that $f_0 \in H^\omega(D_2)$ we obtain the equation

$$\mathrm{Sup}\{|R(f; X, Y)| : f \in H^\omega(D_2)\} = R(f_0; X, Y) = \int_0^1 \int_0^1 f_0(X, Y; x, y)\,dx\,dy.$$

$$(5.1.16)$$

Then, (5.1.16) and (5.1.6) imply (5.1.12).

All the assertions of the theorem follow from (5.1.11) and (5.1.12). Indeed, the first equation of (5.1.11) means that formula (5.1.9), (5.1.10) is the best among all possible cubature formulas for recovery of the functional $\int_0^1 \int_0^1 f(x, y)\,dx\,dy$ in the class $H^{\omega_1 \omega_2}(D_2)$ on information $T(f; X, Y)$, $(X, Y) \in (X, Y)_{0,0}$. The second of the equations of (5.1.11) expresses explicitly the errors. The third of them expresses the extremality of the standard function $f_{12}(X, Y; x, y)$ in $H^{\omega_1 \omega_2}(D_2)$.

From (5.1.12) all the conclusions for the function class $H^\omega(D_2)$ follow. The theorem is proved.

## 5.2   Optimal cubature formulae for the classes $H^{\omega_1\omega_2}(D_2)$ and $H^{\omega}(D_2)$

### Theorem 5.2.1

*Let n and m be fixed positive integers, $v$ and $\mu$ be fixed elements of the set $\{0, 1, 2, 3\}$, $(X, Y) \in (X, Y)_{v\mu}$ and $\omega_1(t)$, $\omega_2(t)$ and $\omega(t)$ be arbitrary moduli of continuity. Then from all cubature formulas for approximate calculation of the integral*

$$Lf = \int_0^1\!\!\int_0^1 f(x, y)\,\mathrm{d}x\,\mathrm{d}y$$

*using information $T(f; X, Y)$, $(X, Y) \in (X, Y)_{v,\mu}$ only, the optimal one in the classes $H^{\omega_1\omega_2}(D_2)$ and $H^{\omega}(D_2)$ is the formula*

$$\int_0^1\!\!\int_0^1 f(x, y)\,\mathrm{d}x\,\mathrm{d}y = Q(f; X_v, Y_\mu) + R(f; X_v, Y_\mu) \tag{5.2.1}$$

*constructed on the basis of formulas (5.1.9), (5.1.10) for $(X, Y) = (X_v, Y_\mu)$, where the mesh $X_v(Y_\mu)$ is defined in $(1.6.5_v)$ (in $(1.6.5_\mu)$ for $n = m$, respectively).*
*In addition,*

$$R_{nm}^{v\mu}(H^{\omega_1\omega_2}(D_2)) = \inf\{R(H^{\omega_1\omega_2}(D_2); X, Y) : (X, Y) \in (X, Y)_{v,\mu}\}$$

$$= R(H^{\omega_1\omega_2}(D_2); X_v, Y_\mu)$$

$$= \int_0^1\!\!\int_0^1 f_{12}(X_v, Y_\mu; x, y)\,\mathrm{d}x\,\mathrm{d}y$$

$$= \int_0^1 \omega_1(u \cdot D(A_v, X_v)_\infty)\,\mathrm{d}u + \int_0^1 \omega_2(u \cdot D(A_\mu, Y_\mu)_\infty)\,\mathrm{d}u \tag{5.2.2}$$

$$R_{nm}^{v\mu}(H^{\omega}(D_2)) = \inf\{R(H^{\omega}(D_2); X, Y) : (X, Y) \in (X, Y)_{v,\mu}\}$$

$$= R(H^{\omega}(D_2); X_v, Y_\mu)$$

$$= \int_0^1 \int_0^1 f_0(X_v, Y_\mu; x, y) dx\, dy$$

$$= \int_0^1 \int_0^1 \omega(\sqrt{u^2 \cdot D^2(A_v, X_v)_\infty + v^2 \cdot D^2(A_\mu, Y_\mu)_\infty})\, du\, dv \qquad (5.2.3)$$

where $D(A, Y)_\infty$ is the uniform deviation of the mesh pair $(A, X)$; the mesh pair $(A_v, X_v)$ is defined in $(1.6.5_v)$, and $(A_\mu, Y_\mu)$ in $(1.6.5_\mu)$ for $n = m$ respectively.

*Extremal for the class $H^{\omega_1 \omega_2}(D_2)$ is the standard function $f_{12}(X_v, Y_\mu; x, y)$, defined in $(3.1.7)$ for $(X, Y) = (X_v, Y_\mu)$; extremal for the class $H^\omega(D_2)$ is the function $f_0(X_v, Y_\mu; x, y)$, defined in $(3.1.8)$, respectively.*

### Proof

In order to find the optimal cubature formula in the class $H^{\omega_1 \omega_2}(D_2)$ (or, in $H^\omega(D_2)$, correspondingly), it is necessary and sufficient to optimize the error of the best cubature formula in the corresponding function class on all possible lattices $(X, Y) \in (X, Y)_{v,\mu}$ of nodes in the unit square $D_2$.

First, we consider the class $H^{\omega_1 \omega_2}(D_2)$. By definition

$$R_{nm}^{v\mu}(H^{\omega_1 \omega_2}(D_2)) = \inf\{R(H^{\omega_1 \omega_2}(D_2); X, Y): (X, Y) \in (X, Y)_{v,\mu}\}. \qquad (5.2.4)$$

According to Theorem 5.1.1

$$\inf\{R(H^{\omega_1 \omega_2}(D_2); X, Y): (X, Y) \in (X, Y)_{v,\mu}\}$$

$$= \inf\{R(f_{12}; X, Y): (X, Y) \in (X, Y)_{v,\mu}\}$$

$$= \inf\left\{\int_0^1 \int_0^1 f_{12}(X, Y; x, y) dx\, dy: (X, Y) \in (X, Y)_{v,\mu}\right\}$$

$$\overset{(5.1.4')}{=} \inf_{(X,Y)\in(X,Y)_{v,\mu}} \left[ \sum_{i=1}^n \int_{b_{i-1}}^n \omega_1(|x - x_i|) dx + \sum_{j=1}^m \int_{e_{j-1}}^{e_j} \omega_2(|y - y_j|) dy \right]$$

$$\overset{(2.1.2)}{=} \inf_{(X,Y)\in(X,Y)_{v,\mu}} [L_n(\omega_1; B, X) + L_m(\omega_2; E, Y)]. \qquad (5.2.5)$$

According to Lemma 3.2.1

$$\inf\{L_n(\omega_1; B, X) + L_m(\omega_2; E, Y): (X, Y)\in(X, Y)_{v,\mu}\}$$

$$= \inf\{L_n(\omega_1; B, X): X\in(X)_v\} + \inf\{L_m(\omega_2; E, Y): Y\in(Y)_\mu\}. \quad (5.2.6)$$

From the other side,

$$\inf\{L_n(\omega_1; B, X): X\in(X)_v\}$$

$$= \inf\{L_n(\omega_1; B, X): (B, X)\in(A, X)_v\}$$

$$\geq \inf\{L_n(\omega_1; A, X): (A, X)\in(A, X)_v\}. \quad (5.2.7)$$

According to the optimization Theorem 2.1.1

$$\inf\{L_n(\omega_1; A_1 X): (A, X): (A, X)\in(A, X)_v\} = L_n(\omega_1; A_v, X_v) \quad (5.2.8)$$

since the function $\omega_1(t)$ is increasing.

Taking into account that the mesh $A_v$ is uniquely determined by the mesh $X_v$ from formulas (1.5.21, (5.2.7) and (5.2.8) we obtain the equality

$$\inf\{L_n(\omega_1; B, X): X\in(X)_v\} = L_n(\omega_1; A_v, X_v). \quad (5.2.9)$$

In an analogous manner, we obtain

$$\inf\{L_m(\omega_2; E, Y): Y\in(Y)_\mu\} = L_n(\omega_2; A_\mu, Y_\mu). \quad (5.2.10)$$

Relations (5.2.4)–(5.2.6) and (5.2.9)–(5.2.10) imply the equality

$$R_{nm}^{v\mu}(H^{\omega_1\omega_2}(D_2)) = L_n(\omega_1; A_v, X_v) + L_m(\omega_2; A_\mu, Y_\mu). \quad (5.2.11)$$

It is essential to observe that (5.2.5) contains the equalities

$$R(H^{\omega_1\omega_2}(D_2); X, Y) = \int_0^1\int_0^1 f_{12}(X, Y; x, y)\mathrm{d}x\,\mathrm{d}y$$

$$= R(f_{12}; X, Y)$$

$$= L_n(\omega_1; B, X) + L_m(\omega_2; E, Y) \quad (5.2.12)$$

where the mesh $B$ is uniquely determined by the mesh $X$ using formulas (1.5.21) and, analogously, $E$ is uniquely determined on $Y$ by (3.1.6).

Observing that $A_v$ is uniquely determined by $X_v$ using (1.5.21), and $A_\mu$ by $Y_\mu$ using (3.1.6), then, by (5.2.12) we may write

$$R(H^{\omega_1\omega_2}(D_2); X_v, Y_\mu) = \int_0^1\int_0^1 f_{12}(X_v, Y_\mu; x, y)\mathrm{d}x\,\mathrm{d}y$$

$$= R(f_{12}; X_v, Y_\mu)$$

$$= L_n(\omega_1; A_v, X_v) + L_m(\omega_2; A_\mu, Y_\mu). \qquad (5.2.13)$$

Then (5.2.11) and (5.2.13) imply equalities (5.2.2) since according to (2.1.3)

$$L_n(\omega_1; A_v, X_v) = \int_0^1 \omega_1(u \cdot D(A_v, X_v)_\infty)\mathrm{d}u$$

and

$$L_m(\omega_2; A_\mu, Y_\mu) = \int_0^1 \omega_2(u \cdot D(A_\mu, Y_\mu)_\infty)\mathrm{d}u.$$

The assertions of the theorem for the function class $H^{\omega_1\omega_2}(D_2)$ are proven, since the equation

$$R_{nm}^{v\mu}(H^{\omega_1\omega_2}(D_2)) = R(H^{\omega_1\omega_2}(D_2); X_v, Y_\mu)$$

by definition means optimality of the mesh $(X_v, Y_\mu) \in (X, Y)_{v,\mu}$ and hence optimality of the cubature formula (5.2.1), and the equations

$$R_{nm}^{v\mu}(H^{\omega_1\omega_2}(D_2)) = R(f_{12}; X_v, Y_\mu) = \int_0^1\int_0^1 f_{12}(X_v, Y_\mu, x, y)\mathrm{d}x\,\mathrm{d}y$$

mean optimality of the standard function $f_{12}(X_v, Y_\mu; x, y)$. They also contain the exact error.

In the same manner we can consider the class $H^\omega(D_2)$. Indeed, by definition

$$R_{nm}^{v\mu}(H^\omega(D_2)) = \inf\{R(H^\omega(D_2); X, Y) : (X, Y) \in (X, Y)_{v,\mu}\} \qquad (5.2.14)$$

According to Theorem 5.1.1

$$\inf\{R(H^\omega(D_2)); X, Y): (X, Y)\in(X, Y)_{v,\mu}\}$$

$$= \inf\{R(f_0; X, Y): (X, Y)\in)_{v,\mu}\}$$

$$= \inf\left\{\int_0^1\int_0^1 f_0(X, Y; x, y)\mathrm{d}x\,\mathrm{d}y: (X, Y)\in(X, Y)_{v,\mu}\right\}$$

$$\overset{(3.1.8)}{=} \inf_{(X,Y)\in(X,Y)_{v,\mu}} \sum_{i=1}^n \int_{b_{i-1}}^{b_i}\left[\sum_{j=1}^m \int_{e_{j-1}}^{e_j} \omega(\sqrt{(x-x_i)^2+(y-y_j)^2})\mathrm{d}y\right]\mathrm{d}x. \quad (5.2.15)$$

Using the possibility of calculating the infimums in arbitrary order (see [6], p 304, formula (D18)), the increasing of the function $\varphi(u)=\omega(\sqrt{u^2+a})$ applying the optimization theorem at first with respect to the variable $x$, and then to $y$, using (5.2.14) and (5.2.15), we get

$$R_{nm}^{v\mu}(H^\omega(D_2)) = \inf_{Y\in(Y)_\mu}\left\{\inf_{X\in(X)_v} \sum_{i=1}^n \int_{b_{i-1}}^{b_i}\left[\sum_{j=1}^n \int_{e_{j-1}}^{e_j} \omega(\sqrt{(x-x_i)^2+(y-y_j)^2})\mathrm{d}y\right]\mathrm{d}x\right\}$$

$$= \inf_{Y\in(Y)_\mu} \int_0^1\left[\sum_{j=1}^m \int_{e_{j-1}}^{e_j} \omega(\sqrt{u^2\cdot D^2(A_v, X_v)_\infty+(y-y_j)^2})\mathrm{d}y\right]\mathrm{d}u$$

$$= \int_0^1\int_0^1 \omega(\sqrt{u^2\cdot D^2(A_v, X_v)_\infty+v^2\cdot D^2(A_\mu, Y_\mu)_\infty})\mathrm{d}v\,\mathrm{d}u. \quad (5.2.16)$$

(It can easily be shown that the exchange $\inf\int_0^1\to\int_0^1\inf$ used in the last equality is justified.)

The identity

$$\int_0^1\int_0^1 f_0(X_v, Y_\mu; x, y)\mathrm{d}x\,\mathrm{d}y = \int_0^1\int_0^1 \omega(\sqrt{u^2\cdot D^2(A_v, X_v)_\infty+v^2\cdot D^2(A_\mu, Y_\mu)_\infty})\mathrm{d}v\,\mathrm{d}u$$

can be verified directly. This equality and (5.2.16) imply (5.2.9) and thus the theorem is proved.

The case $(v, \mu)=(0,0)$ of Theorem 5.2.1 is due to Korneichuk. More specifically, the following corollary is true.

**Corollary 5.2.1** (*Korneichuk* [19].) *Out of all cubature formulas for approximate calculation of the integral $\int_0^1\int_0^1 f(x,y)\mathrm{d}x\,\mathrm{d}y = Lf$ usually the information $T(f;X,Y)$, $(X,Y)\in(X,Y)_{0,0}$, optimal in the classes $H^{\omega_1\omega_2}(D_2)$ and $H^\omega(D_2)$ is the formula*

$$\int_0^1\int_0^1 f(x,y)\mathrm{d}x\,\mathrm{d}y = \frac{1}{nm}\sum_{i=1}^n\sum_{j=1}^m f\left(\frac{2i-1}{2n},\frac{2j-1}{2m}\right) + R(f;X_0,Y_0). \quad (5.2.17)$$

*with*

$$R_{nm}^{00}(H^{\omega_1\omega_2}(D_2)) = \int_0^1 \omega_1\left(\frac{u}{2n}\right)\mathrm{d}u + \int_0^1 \omega_2\left(\frac{u}{2n}\right)\mathrm{d}u \quad (5.2.18)$$

*and*

$$R_{nm}^{00}(H^\omega(D_2)) = \int_0^1\int_0^1 \omega\left(\frac{1}{2}\sqrt{\frac{u^2}{n^2}+\frac{v^2}{m^2}}\right)\mathrm{d}u\,\mathrm{d}v. \quad (5.2.19)$$

Extremal in $H^{\omega_1\omega_2}(D_2)$ is the standard function $f_{12}(X_0,Y_0;x,y)$ and the function $f_0(X_0,Y_0;x,y)$ in the class $H^\omega(D_2)$, correspondingly.

For the applications it is essential to have an explicit expression for $Q(f;X_v,Y_\mu)$. That is why we give a table of explicit expressions for $Q(f;X_v,Y_\mu)$ for various $(v,\mu)$.

In table 5.2.1 we give the assertions of Theorem 5.2.1 for various $(v,\mu)$. Each row of this table gives an assertion of the theorem for given $v,\mu$; restriction, if any, for $n$ and $m$; $Q(f;X_v,Y_\mu)$; the error $R_{nm}^{v\mu}(H^{\omega_1\omega_2}(D_2))$ and the extremal function for the class $H^{\omega_1\omega_2}(D_2)$ and, finally, the error and the extremal function for the class $H^\omega(D_2)$.

$$Q(f;X_0,Y_0) = \frac{1}{nm}\sum_{i=1}^n\sum_{j=1}^m f\left(\frac{2i-1}{2n},\frac{2j-1}{2m}\right)$$

$$Q(f;X_1,Y_0) = \frac{1}{(2n-1)m}\sum_{j=1}^m f\left(0,\frac{2j-1}{2m}\right)$$

$$+\frac{2}{(2n-1)m}\sum_{i=2}^n\sum_{j=1}^m f\left(\frac{2i-2}{2n-1},\frac{2j-1}{2m}\right)$$

$$Q(f;X_2,Y_0) = \frac{1}{(2n-2)m} \sum_{j=1}^{m} \left[ f\left(0,\frac{2j-1}{2m}\right) + f\left(1,\frac{2j-1}{2m}\right) \right]$$

$$+ \frac{1}{(n-1)m} \sum_{i=2}^{n-1} \sum_{j=1}^{m} f\left(\frac{i-1}{n-1},\frac{2j-1}{2m}\right)$$

$$Q(f;X_3,Y_0) = \frac{1}{(2n-1)m} \sum_{j=1}^{m} f\left(1,\frac{2j-1}{2m}\right)$$

$$+ \frac{2}{(2n-1)m} \sum_{i=1}^{n-1} \sum_{j=1}^{m} f\left(\frac{2i-1}{2n-1},\frac{2j-1}{2m}\right)$$

$$Q(f;X_0,Y_1) = \frac{1}{n(2m-1)} \sum_{i=1}^{n} f\left(\frac{2i-1}{2n},0\right)$$

$$+ \frac{2}{n(2m-1)} \sum_{i=1}^{n} \sum_{j=2}^{m} f\left(\frac{2i-1}{2n},\frac{2j-2}{2m-1}\right)$$

$$Q(f;X_1,Y_1) = \frac{f(0,0)}{(2n-1)(2m-1)} + \frac{2}{(2n-1)(2m-1)}$$

$$\times \left[ \sum_{i=2}^{n} f\left(\frac{2i-2}{2n-1},0\right) + \sum_{j=2}^{m} f\left(0,\frac{2j-2}{2m-1}\right) \right]$$

$$+ \frac{4}{(2n-1)(2m-1)} \sum_{i=2}^{n} \sum_{j=2}^{m} f\left(\frac{2i-2}{2n-1},\frac{2j-2}{2m-1}\right)$$

$$Q(f;X_2,Y_1) = \frac{f(0,0)+f(1,0)}{(2n-2)(2m-1)} + \frac{1}{(n-1)(2m-1)} \sum_{i=2}^{n-1} f\left(\frac{i-1}{n-1},0\right)$$

$$+ \frac{1}{(n-1)(2m-1)} \sum_{j=2}^{m} \left[ f\left(0,\frac{2j-2}{2m-1}\right) + f\left(1,\frac{2j-2}{2m-1}\right) \right]$$

$$+ \frac{2}{(n-1)(2m-1)} \sum_{i=2}^{n-1} \sum_{j=2}^{m} f\left(\frac{i-1}{n-1},\frac{2j-2}{2m-1}\right)$$

*(continued on pp 146–9)*

**Table 5.2.1**

| $v$ | $\mu$ | $n$ | $m$ | $Q(f;X_v,Y_\mu)$ | $H^{\omega_1\omega_2}(D_2)$<br>$R_{nm}^{v\mu}(H^{\omega_1\omega_2}(D_2))$ |
|---|---|---|---|---|---|
| 0 | 0 | | | $Q(f;X_0,Y_0)$ | $\displaystyle\int_0^1 \omega_1\!\left(\frac{u}{2n}\right)du + \int_0^1 \omega_2\!\left(\frac{u}{2m}\right)du$ |
| 1 | 0 | | | $Q(f;X_1,Y_0)$ | $\displaystyle\int_0^1 \omega_1\!\left(\frac{u}{2n-1}\right)du + \int_0^1 \omega_2\!\left(\frac{u}{2m}\right)du$ |
| 2 | 0 | $n>1$ | | $Q(f;X_2,Y_0)$ | $\displaystyle\int_0^1 \omega_1\!\left(\frac{u}{2n-2}\right)du + \int_0^1 \omega_2\!\left(\frac{u}{2m}\right)du$ |
| 3 | 0 | | | $Q(f;X_3,Y_0)$ | $\displaystyle\int_0^1 \omega_1\!\left(\frac{u}{2n-1}\right)du + \int_0^1 \omega_2\!\left(\frac{u}{2m}\right)du$ |
| 0 | 1 | | | $Q(f;X_0,Y_1)$ | $\displaystyle\int_0^1 \omega_1\!\left(\frac{u}{2n}\right)du + \int_0^1 \omega_2\!\left(\frac{u}{2m-1}\right)du$ |
| 1 | 1 | | | $Q(f;X_1,Y_1)$ | $\displaystyle\int_0^1 \omega_1\!\left(\frac{u}{2n-1}\right)du + \int_0^1 \omega_2\!\left(\frac{u}{2m-1}\right)du$ |
| 2 | 1 | $n>1$ | | $Q(f;X_2,Y_1)$ | $\displaystyle\int_0^1 \omega_1\!\left(\frac{u}{2n-2}\right)du + \int_0^1 \omega_2\!\left(\frac{u}{2m-1}\right)du$ |
| 3 | 1 | | | $Q(f;X_3,Y_1)$ | $\displaystyle\int_0^1 \omega_1\!\left(\frac{u}{2n-1}\right)du + \int_0^1 \omega_2\!\left(\frac{u}{2m-1}\right)du$ |
| 0 | 2 | | $m>1$ | $Q(f;X_0,Y_2)$ | $\displaystyle\int_0^1 \omega_1\!\left(\frac{u}{2n}\right)du + \int_0^1 \omega_2\!\left(\frac{u}{2m-2}\right)du$ |
| 1 | 2 | | $m>1$ | $Q(f;X_1,Y_2)$ | $\displaystyle\int_0^1 \omega_1\!\left(\frac{u}{2n-1}\right)du + \int_0^1 \omega_2\!\left(\frac{u}{2m-2}\right)du$ |
| 2 | 2 | $n>1$ | $m>1$ | $Q(f;X_2,Y_2)$ | $\displaystyle\int_0^1 \omega_1\!\left(\frac{u}{2n-2}\right)du + \int_0^1 \omega_2\!\left(\frac{u}{2m-2}\right)du$ |

| | $H^\omega(D_2)$ | |
| --- | --- | --- |
| $f_{12}(X_\nu Y_\mu; x, y)$ | $R_{nm}^{\nu\mu}(H^\omega(D_2))$ | $f_0(X_\nu, Y_\nu; x, y)$ |
| $f_{12}(X_0, Y_0; x, y)$ | $\displaystyle\int_0^1\!\!\int_0^1 \omega\!\left(\sqrt{\frac{u^2}{(2n)^2} + \frac{v^2}{(2m)^2}}\right) du\, dv$ | $f_0(X_0, Y_0; x, y)$ |
| $f_{12}(X_1, Y_0; x, y)$ | $\displaystyle\int_0^1\!\!\int_0^1 \omega\!\left(\sqrt{\frac{u^2}{(2n-1)^2} + \frac{v^2}{(2m)^2}}\right) du\, dv$ | $f_0(X_1, Y_0; x, y)$ |
| $f_{12}(X_2, Y_0; x, y)$ | $\displaystyle\int_0^1\!\!\int_0^1 \omega\!\left(\sqrt{\frac{u^2}{(2n-2)^2} + \frac{v^2}{(2m)^2}}\right) du\, dv$ | $f_0(X_2, Y_0; x, y)$ |
| $f_{12}(X_3, Y_0; x, y)$ | $\displaystyle\int_0^1\!\!\int_0^1 \omega\!\left(\sqrt{\frac{u^2}{(2n-1)^2} + \frac{v^2}{(2m)^2}}\right) du\, dv$ | $f_0(X_3, Y_0; x, y)$ |
| $f_{12}(X_0, Y_1; x, y)$ | $\displaystyle\int_0^1\!\!\int_0^1 \omega\!\left(\sqrt{\frac{u^2}{(2n)^2} + \frac{v^2}{(2m-1)^2}}\right) du\, dv$ | $f_0(X_0, Y_1; x, y)$ |
| $f_{12}(X_1, Y_1; x, y)$ | $\displaystyle\int_0^1\!\!\int_0^1 \omega\!\left(\sqrt{\frac{u^2}{(2n-1)^2} + \frac{v^2}{(2m-1)^2}}\right) du\, dv$ | $f_0(X_1, Y_1; x, y)$ |
| $f_{12}(X_2, Y_1; x, y)$ | $\displaystyle\int_0^1\!\!\int_0^1 \omega\!\left(\sqrt{\frac{u^2}{(2n-2)^2} + \frac{v^2}{(2m-1)^2}}\right) du\, dv$ | $f_0(X_2, Y_1; x, y)$ |
| $f_{12}(X_3, Y_1; x, y)$ | $\displaystyle\int_0^1\!\!\int_0^1 \omega\!\left(\sqrt{\frac{u^2}{(2n-1)^2} + \frac{v^2}{(2m-1)^2}}\right) du\, dv$ | $f_0(X_3, Y_1; x, y)$ |
| $f_{12}(X_0, Y_2; x, y)$ | $\displaystyle\int_0^1\!\!\int_0^1 \omega\!\left(\sqrt{\frac{u^2}{(2n)^2} + \frac{v^2}{(2m-2)^2}}\right) du\, dv$ | $f_0(X_0, Y_2; x, y)$ |
| $f_{12}(X_1, Y_2; x, y)$ | $\displaystyle\int_0^1\!\!\int_0^1 \omega\!\left(\sqrt{\frac{u^2}{(2n-1)^2} + \frac{v^2}{(2m-2)^2}}\right) du\, dv$ | $f_0(X_1, Y_2; x, y)$ |
| $f_{12}(X_2, Y_2; x, y)$ | $\displaystyle\int_0^1\!\!\int_0^1 \omega\!\left(\sqrt{\frac{u^2}{(2n-2)^2} + \frac{v^2}{(2m-2)^2}}\right) du\, dv$ | $f_0(X_2, Y_2; x, y)$ |

*(continued)*

**Table 5.2.1**　(*continued*)

| $\nu$ | $\mu$ | $n$ | $m$ | $Q(f;X_\nu,Y_\mu)$ | $H^{\omega_1\omega_2}(D_2)$ $R^{\nu\mu}_{nm}(H^{\omega_1\omega_2}(D_2))$ |
|---|---|---|---|---|---|
| 3 | 2 | | $m>1$ | $Q(f;X_3,Y_2)$ | $\displaystyle\int_0^1 \omega_1\left(\frac{u}{2n-1}\right)du + \int_0^1 \omega_2\left(\frac{u}{2m-2}\right)du$ |
| 0 | 3 | | | $Q(f;X_0,Y_3)$ | $\displaystyle\int_0^1 \omega_1\left(\frac{u}{2n}\right)du + \int_0^1 \omega_2\left(\frac{u}{2m-1}\right)du$ |
| 1 | 3 | | | $Q(f;X_1,Y_3)$ | $\displaystyle\int_0^1 \omega_1\left(\frac{u}{2n-1}\right)du + \int_0^1 \omega_2\left(\frac{u}{2m-1}\right)du$ |
| 2 | 3 | $n>1$ | | $Q(f;X_2,Y_3)$ | $\displaystyle\int_0^1 \omega_1\left(\frac{u}{2n-2}\right)du + \int_0^1 \omega_2\left(\frac{u}{2m-1}\right)du$ |
| 3 | 3 | | | $Q(f;X_3,Y_3)$ | $\displaystyle\int_0^1 \omega_1\left(\frac{u}{2n-1}\right)du + \int_0^1 \omega_2\left(\frac{u}{2m-1}\right)du$ |

$$Q(f;X_3,Y_1) = \frac{f(1,0)}{(2n-1)(2m-1)} + \frac{2}{(2n-1)(2m-1)}$$

$$\times \left[\sum_{i=1}^{n-1} f\left(\frac{2i-1}{2n-1},0\right) + \sum_{j=2}^{m} f\left(1,\frac{2j-2}{2m-1}\right)\right]$$

$$+ \frac{4}{(2n-1)(2m-1)} \sum_{i=1}^{n-1}\sum_{j=2}^{m} f\left(\frac{2i-1}{2n-1},\frac{2j-2}{2m-1}\right)$$

$$Q(f;X_0,Y_2) = \frac{1}{n(2m-2)} \sum_{i=1}^{n}\left[f\left(\frac{2i-1}{2n},0\right) + f\left(\frac{2i-1}{2n},1\right)\right]$$

$$+ \frac{1}{n(m-1)} \sum_{i=1}^{n}\sum_{j=2}^{m-1} f\left(\frac{2i-1}{2n},\frac{j-1}{m-1}\right)$$

$$Q(f;X_1,Y_2) = \frac{f(0,0)+f(0,1)}{(2n-1)(2m-2)} + \frac{1}{(2n-1)(m-1)} \sum_{j=2}^{m-1} f\left(0,\frac{j-1}{m-1}\right)$$

| $f_{12}(X_\nu Y_\mu; x, y)$ | $H^\omega(D_2)$ | |
| --- | --- | --- |
| | $R_{nm}^{\nu\mu}(H^\omega(D_2))$ | $f_0(X_\nu, Y_\nu; x, y)$ |
| $f_{12}(X_3, Y_2; x, y)$ | $\displaystyle\int_0^1\int_0^1 \omega\left(\sqrt{\frac{u^2}{(2n-1)^2} + \frac{v^2}{(2m-2)^2}}\right) du\, dv$ | $f_0(X_3, Y_2; x, y)$ |
| $f_{12}(X_0, Y_3; x, y)$ | $\displaystyle\int_0^1\int_0^1 \omega\left(\sqrt{\frac{u^2}{(2n)^2} + \frac{v^2}{(2m-1)^2}}\right) du\, dv$ | $f_0(X_0, Y_3; x, y)$ |
| $f_{12}(X_1, Y_3; x, y)$ | $\displaystyle\int_0^1\int_0^1 \omega\left(\sqrt{\frac{u^2}{(2n-1)^2} + \frac{v^2}{(2m-1)^2}}\right) du\, dv$ | $f_0(X_1, Y_3; x, y)$ |
| $f_{12}(X_2, Y_3; x, y)$ | $\displaystyle\int_0^1\int_0^1 \omega\left(\sqrt{\frac{u^2}{(2n-2)^2} + \frac{v^2}{(2m-1)^2}}\right) du\, dv$ | $f_0(X_2, Y_3; x, y)$ |
| $f_{12}(X_3, Y_3; x, y)$ | $\displaystyle\int_0^1\int_0^1 \omega\left(\sqrt{\frac{u^2}{(2n-1)^2} + \frac{v^2}{(2m-1)^2}}\right) du\, dv$ | $f_0(X_3, Y_3; x, y)$ |

$$+ \frac{1}{(2n-1)(m-1)} \sum_{i=2}^{n} \left[ f\left(\frac{2i-2}{2n-1}, 0\right) + f\left(\frac{2i-2}{2n-1}, 1\right) \right]$$

$$+ \frac{2}{(2n-1)(m-1)} \sum_{i=2}^{n} \sum_{j=2}^{m-1} f\left(\frac{2i-2}{2n-1}, \frac{j-1}{m-1}\right)$$

$$Q(f; X_2, Y_2) = \frac{1}{4(n-1)(m-1)} \left[ f(0,0) + f(0,1) + f(1,0) + f(1,1) \right]$$

$$+ \frac{1}{2(n-1)(m-1)} \sum_{i=2}^{n-1} \left[ f\left(\frac{i-1}{n-1}, 0\right) + f\left(\frac{i-1}{n-1}, 1\right) \right]$$

$$+ \frac{1}{2(n-1)(m-1)} \sum_{j=2}^{m-1} \left[ f\left(0, \frac{j-1}{m-1}\right) + f\left(1, \frac{j-1}{m-1}\right) \right]$$

$$+ \frac{1}{(n-1)(m-1)} \sum_{i=2}^{n-1} \sum_{j=2}^{m-1} f\left(\frac{i-1}{n-1}, \frac{j-1}{m-1}\right)$$

*(continued on p 148)*

$$Q(f;X_3,Y_2) = \frac{f(1,0)+f(1,1)}{(2n-1)(2m-2)} + \frac{1}{(2n-1)(m-1)} \sum_{j=2}^{m-1} f\left(1,\frac{j-1}{m-1}\right)$$

$$+ \frac{1}{(2n-1)(m-1)} \sum_{i=1}^{n-1} \left[ f\left(\frac{2i-1}{2n-1},0\right) + f\left(\frac{2i-1}{2n-1},1\right) \right]$$

$$+ \frac{2}{(2n-1)(m-1)} \sum_{i=1}^{n-1} \sum_{j=2}^{m-1} f\left(\frac{2i-1}{2n-1},\frac{j-1}{m-1}\right)$$

$$Q(f;X_0,Y_3) = \frac{1}{n(2m-1)} \sum_{i=1}^{n} f\left(\frac{2i-1}{2n},1\right)$$

$$+ \frac{2}{n(2m-1)} \sum_{i=1}^{n} \sum_{j=1}^{m-1} f\left(\frac{2i-1}{2n},\frac{2j-1}{2m-1}\right)$$

$$Q(f;X_1,Y_3) = \frac{f(0,1)}{(2n-1)(2m-1)} + \frac{2}{(2n-1)(2m-1)}$$

$$\times \left[ \sum_{i=2}^{n} f\left(\frac{2i-2}{2n-1},1\right) + \sum_{j=1}^{m-1} f\left(0,\frac{2j-1}{2m-1}\right) \right]$$

$$+ \frac{4}{(2n-1)(2m-1)} \sum_{i=2}^{n} \sum_{j=1}^{m-1} f\left(\frac{2i-2}{2n-1},\frac{2j-1}{2m-1}\right)$$

$$Q(f;X_2,Y_3) = \frac{f(0,1)+(1,1)}{(2n-2)(2m-1)} + \frac{1}{(n-1)(2m-1)} \sum_{i=2}^{n-1} f\left(\frac{i-1}{n-1},1\right)$$

$$+ \frac{1}{(n-1)(2m-1)} \sum_{j=1}^{m-1} \left[ f\left(0,\frac{2j-1}{2m-1}\right) + f\left(1,\frac{2j-1}{2m-1}\right) \right]$$

$$+ \frac{2}{(n-1)(2m-1)} \sum_{i=2}^{n-1} \sum_{j=1}^{m-1} f\left(\frac{i-1}{n-1},\frac{2j-1}{2n-1}\right)$$

$$Q(f; X_3, Y_3) = \frac{f(1,1)}{(2n-1)(2m-1)} + \frac{2}{(2n-1)(2m-1)}$$

$$\times \left[ \sum_{i=1}^{n-1} f\left(\frac{2i-1}{2n-1}, 1\right) + \sum_{j=1}^{m-1} f\left(1, \frac{2j-1}{2m-1}\right) \right]$$

$$+ \frac{4}{(2n-1)(2m-1)} \sum_{i=1}^{n-1} \sum_{j=1}^{m-1} f\left(\frac{2i-1}{2n-1}, \frac{2j-1}{2m-1}\right).$$

# 6

# Approximation of functions by rational quasi-splines

The quasi-splines of degree $r$ defined in (1.5.25), used in the previous sections, in general have (along with all their derivatives) discontinuities in the nodes of the mesh (1.5.21), or, as it is usually said, they have maximal defect (equal to $r + 1$) in these nodes. In the present chapter we consider the approximate properties of another class of quasi-splines. This is the class of the rational quasi-splines, the defect of which is not maximal (it could be made equal to zero).

## 6.1   Definition of the basic quasi-splines. Examples

Let $A$ be the set of all finite functions $f : \mathbb{R} \to \mathbb{R}$ (with carriers in the segment $[-1, 1]$) which are even, increasing over the segment $[-1, 0]$ such that the area on the segment $[-1, 1]$ formed by the graph of the function is non-zero. To be specific, to $A$ belong all functions $f : \mathbb{R} \to \mathbb{R}$ having the properties:

(a)  $f(x) = 0 \qquad \forall x \in \mathbb{R} \backslash (-1, 1)$;

(b)  $f(-x) = f(x) \qquad \forall x \in \mathbb{R}$;

(c)  $-1 \leq x_1 < x_2 \leq 0 \Rightarrow f(x_1) \leq f(x_2)$;

(d)  $\int_{-1}^{1} f(t)\mathrm{d}t \neq 0$.

Further, we use the notation: $AC(-1, 1) = A \cap C(-1, 1)$ and $AC[-1, 1] = A \cap C[-1, 1]$.

For each function $k(x) \in A$ we define [36] the basic quasi-spline $k_m(x)$ of order $m$ $(m \in \mathbb{N}_0)$ by the formulas:

$$k_0(x) = \frac{k(x)}{\int_{-1}^{1} k(t)\mathrm{d}t}, \qquad x \in \mathbb{R} \tag{6.1.1}$$

$$k_m(x) = \int_{2x-1}^{2x+1} k_{m-1}(t)\,dt, \quad x \in \mathbb{R}, \quad m \in \mathbb{N}. \tag{6.1.2}$$

**Example 6.1.1**   The function (figure 6.1.1)

$$\lambda_0(x) = \begin{cases} 1, & |x| < 1/2 \\ 1/2, & |x| = 1/2 \\ 0, & |x| > 1/2 \end{cases} \tag{6.1.3}$$

belongs to the set $A$. By $\lambda_m(x)$ we denote its basic quasi-spline. For example, (see figures 6.1.2 and 6.1.3):

$$\lambda_1(x) = \begin{cases} 1, & |x| \leq 1/4 \\ (3/2) - 2\cdot|x|, & 1/4 \leq |x| \leq 3/4 \\ 0, & |x| \geq 3/4 \end{cases} \tag{6.1.4}$$

$$\lambda_2(x) = \begin{cases} 1, & |x| \leq 1/8 \\ 1 - 4((1/8) - |x|)^2, & 1/8 \leq |x| \leq 3/8 \\ (3/2) - 2\cdot|x|, & 3/8 \leq |x| \leq 5/8 \\ 4((7/8) - |x|)^2, & 5/8 \leq |x| \leq 7/8 \\ 0, & |x| \geq 7/8 \end{cases} \tag{6.1.5}$$

and so on.

Relations (6.1.3), (6.1.4) and (6.1.5) imply the inequalities

$$\lambda_0(x) \geq \lambda_1(x) \geq \lambda_2(x), \quad 0 \leq x \leq 1/2. \tag{6.1.6}$$

**Example 6.1.2**   (Rvachev [37], Konovalov [38].) The function (figure 6.1.4)

$$\delta_0(x) = \begin{cases} 1/2, & |x| < 1 \\ 0, & |x| \geq 1 \end{cases} \tag{6.1.7}$$

belongs to $AC(-1, 1)$. Its basic quasi-spline of order $m$ is denoted by $\delta_m(x)$. For example, (figures 6.1.5 and 6.1.6):

$$\delta_1(x) = \begin{cases} 1 - |x|, & |x| \leq 1 \\ 0, & |x| \geq 1 \end{cases} \tag{6.1.8}$$

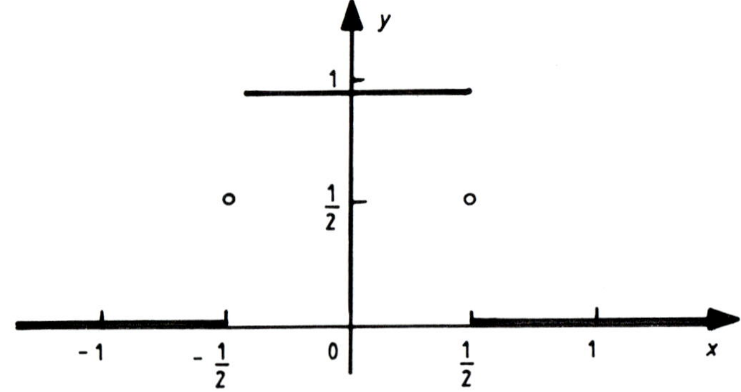

**Figure 6.1.1**   Graph of the basic quasi-spline $\lambda_0(x)$.

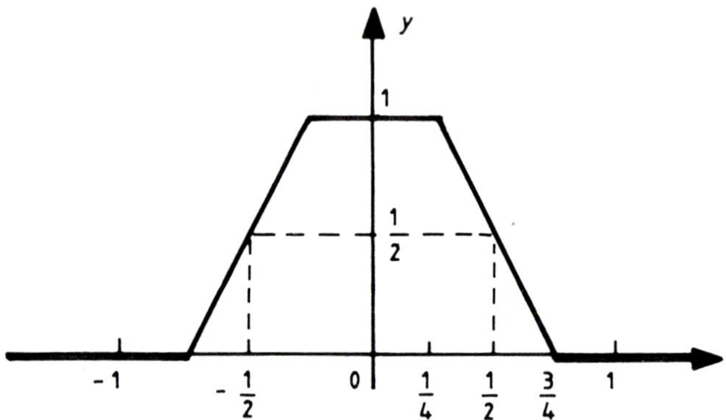

**Figure 6.1.2**   Graph of the basic quasi-spline $\lambda_1(x)$.

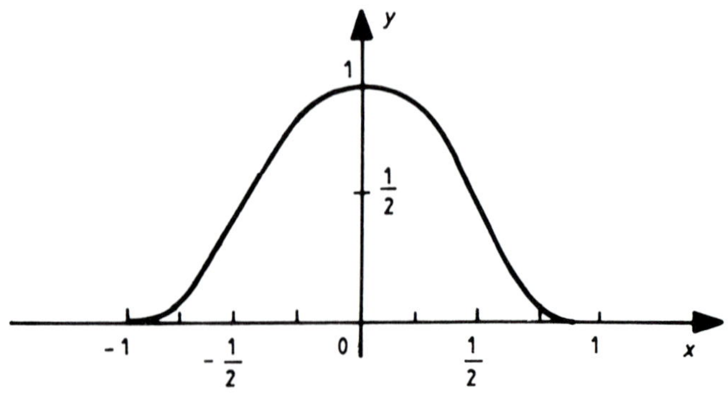

**Figure 6.1.3**   Graph of the basic quasi-spline $\lambda_2(x)$.

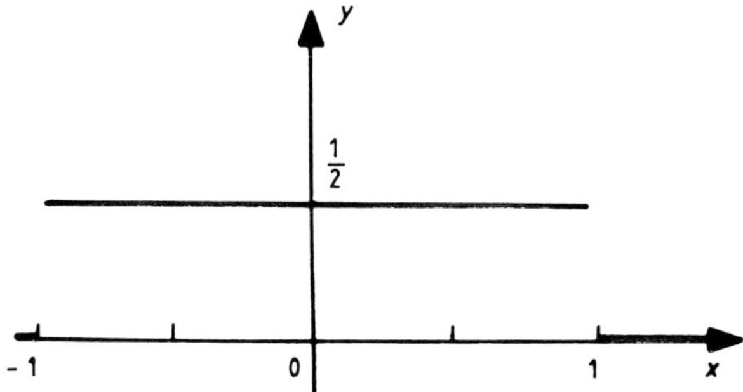

**Figure 6.1.4**  Graph of the basic quasi-spline $\delta_0(x)$.

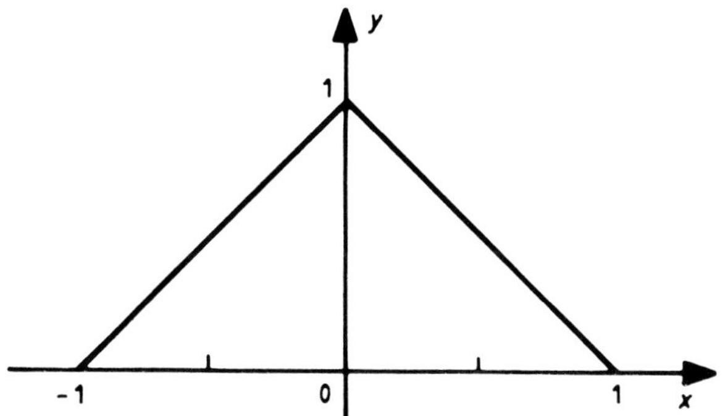

**Figure 6.1.5**  Graph of the basic quasi-spline $\delta_1(x)$.

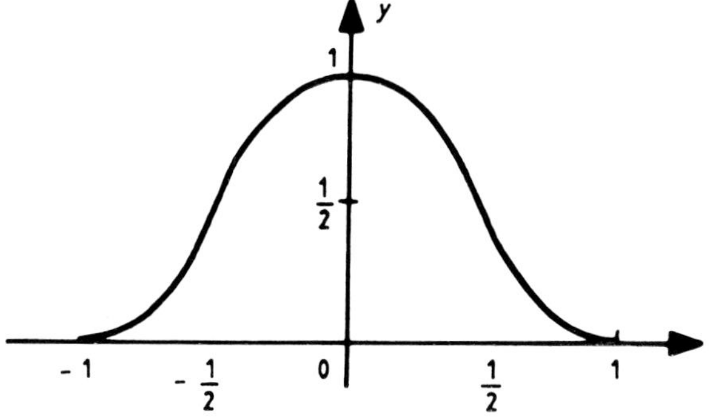

**Figure 6.1.6**  Graph of the basic quasi-spline $\delta_2(x)$.

$$\delta_2(x) = \begin{cases} 1 - 2x^2, & |x| \leq 1/2 \\ 2(1 - |x|)^2, & 1/2 \leq |x| \leq 1 \\ 0, & |x| \geq 1/2 \end{cases} \qquad (6.1.9)$$

and so on.

Relations (6.1.7)–(6.1.9) imply the inequalities

$$\delta_0(x) \leq \delta_1(x) \leq \delta_2(x), \quad 0 \leq x \leq 1/2. \qquad (6.1.10)$$

**Example 6.1.3**   The function

$$\cos_0 x = \begin{cases} (\pi/4)\cos(\pi x/2), & |x| \leq 1 \\ 0, & |x| \geq 1 \end{cases} \qquad (6.1.11)$$

belongs to the set $AC[-1, 1]$.

Its basic quasi-spline of order $m$ is denoted by $\cos_m x$. For example,

$$\cos_1 x = \begin{cases} \cos^2(\pi x/2), & |x| \leq 1 \\ 0, & |x| \geq 1 \end{cases} . \qquad (6.1.12)$$

**Example 6.1.4**   Storchai and Ligun [39], for each function $\varphi(t)$, summable, even, non-negative and decreasing over the segment $[0, 1]$, they define

$$\rho(x) = \int_{-1}^{x} \varphi(t)\,dt, \quad x \in [-1, 1] \qquad (6.1.13)$$

and as a basic quasi-spline they use the function

$$\rho_1(x) = \frac{\rho(1 - 2x)}{\rho(1)}, \quad x \in [0, 1]. \qquad (6.1.14)$$

In [39] it is observed that for $\varphi(x) = (1 - x^2)^r$, $r \in \mathbb{N}$, the construction (6.1.13), (6.1.14) gives the basic quasi-spline of degree $2r + 1$, considered by Velikin [40].

Relations (6.1.1), (6.1.2) and (6.1.13), (6.1.14) imply the equation

$$\rho_1(x) = \varphi_1(x), \quad x \in [0, 1] \qquad (6.1.15)$$

where $\varphi_1(x)$ is the basic quasi-spline of order 1 for the function $\varphi(x)$.

## 6.2  Elementary properties of the basic quasi-splines

In this section we study the elementary properties of the basic quasi-splines under the assumption that the generating function $k(x)$ belongs to the set $A$. Further, writing $k_m(x)$ we keep in mind the quasi-spline of order $m$, defined in (6.1.1) and (6.1.2).

1* $k_m(x) \in R(\mathbb{R})$, i.e. the function $k_m(x)$ is Riemann-integrable on the whole real axis $\mathbb{R}$.

2* $k_m(x) = 0$ for every $x \in \mathbb{R} \backslash (-1, 1)$. Although we must consider the two cases: $x \leq -1$ and $x \geq 1$, the considerations in both cases are analogous, and we may consider only one of them, for example the first one.

If $x \leq -1$, then $k(x) = 0$ since $k(x) \in A$. Now the assertion $k_0(x) = 0$ for $x \leq -1$ follows from (6.1.1), i.e. this property is true for $m = 0$. Further, let this property be true for some $m \in \mathbb{N}_0$, i.e. let $k_m(x) = 0$ for $x \leq -1$.

Then, from $x \leq -1$ we have $2x + 1 \leq -1$ and from the induction assumption and from (6.1.2) we get

$$k_{m+1}(x) = \int_{2x-1}^{2x+1} k_m(t)\,dt = \int_{2x-1}^{2x+1} 0\,dt = 0, \quad x \leq -1.$$

The property is established.

3* $k_m(x) \geq 0$ for $x \in \mathbb{R}$. Indeed, from $k_m(x) \in A$ and (6.1.1) it follows $k_0(x) \geq 0$. Let $k_m(x) \geq 0$ for each $x \in \mathbb{R}$ and for some $m \in \mathbb{N}_0$. Then from $2x - 1 < 2x + 1$ and (6.1.2) it follows that $k_{m+1}(x) \geq 0$ for $x \in \mathbb{R}$.

4* $k_m(-x) = k_m(x)$ for each $x \in \mathbb{R}$. for $m = 0$ the property follows from the condition $(b)$ for set $A$. Let $k_m(-x) = k_m(x)$ for each $x \in \mathbb{R}$ and for some $m \in \mathbb{N}_0$. Then

$$k_{m+1}(-x) = \int_{-2x-1}^{-2x+1} k_m(t)\,dt = -\int_{2x+1}^{2x-1} k_m(-\tau)\,d\tau = \int_{2x-1}^{2x+1} k_n(\tau)\,d\tau = k_{m+1}(x)$$

The property is proved.

5* $-1 \leq x_1 < x_2 \leq 0 \Rightarrow k_m(x_1) \leq k_m(x_2)$. For $m = 0$ the property follows from (6.1.1) and condition $(c)$ for the set $A$. Let it be true for some $m \in \mathbb{N}_0$. Since

$$-1 \leq x_1 < x_2 \leq 0 \Rightarrow \begin{array}{c} 2x_1 - 1 < 2x_2 - 1 \leq -1 \\ -1 \leq 2x_1 + 1 < 2x_2 + 1 \leq 1 \end{array} \tag{6.2.1}$$

then for $m + 1$ of 2* and 3* and from (6.1.2) it follows that

$$k_{m+1}(x_2) - k_{m+1}(x_1) = \int_{-1}^{2x_2+1} k_m(t)\,dt - \int_{-1}^{2x_1+1} k_m(t)\,dt = \int_{2x_1+1}^{2x_2+1} k_m(t)\,dt \geq 0$$

and thus the property is proved.

$6^*$  $k_m(0) = \int_{-1}^{1} k_{m-1}(t)dt = 1$,   $m \in \mathbb{N}$. For $m = 1$, the property follows from (6.1.1) and (6.1.2).

$$k_1(0) = \int_{-1}^{1} k_0(\tau)d\tau = \frac{\int_{-1}^{1} k(\tau)d\tau}{\int_{-1}^{1} k(t)dt} = 1.$$

Let it be true for some $m \in \mathbb{N}$. Then, from (6.1.2) by interchanging the order of the integration, we obtain

$$k_{m+1}(0) = \int_{-1}^{1} k_m(t)dt = \int_{-1}^{1} \int_{2t-1}^{2t+1} k_{m-1}(u)du\, dt$$

$$= \int_{-1}^{1} \int_{(u-1)/2}^{(u+1)/2} k_{m-1}(u)dt\, du = \int_{-1}^{1} k_{m-1}(u)du = k_m(0) = 1.$$

The proof is accomplished.

$7^*$  $k_m(x) \in C^{m-1}(\mathbb{R})$ for each $m \in \mathbb{N}$. If $k(x) \in AC[-1,1]$, then $k_m(x) \in C^m(\mathbb{R})$ for every $m \in \mathbb{N}_0$. The assertion follows from (6.1.2) and from the known properties of the integral as a function of its upper limit.

The properties $2^*$, $4^*$, $5^*$ and $6^*$ imply

$8^*$  $k_m(x) \in A$, $m \in \mathbb{N}_0$.

$9^*$  If $m \in \mathbb{N}$, $s = 1, 2, \ldots, m-1$, then for each $x \in \mathbb{R}$

$$k_m^{(s)}(x) = 2^{\frac{s(+1)}{2}} \sum_{i_1=0}^{1} \cdots \sum_{i_s=0}^{1} (-1)^{i_1 + \ldots + i_s} k_{m-s}(2^s x + A_i(s)) \tag{6.2.2}$$

with $i = (i_1, i_2, \ldots, i_s)$ and

$$A_i(s) = \sum_{j=1}^{s} (-1)^{i_j} 2^{s-j}, \quad s \in \mathbb{N}. \tag{6.2.3}$$

If $k(x) \in AC[-1,1]$, then (6.2.2) is true for $s = m$ also.
Indeed, from (6.1.2) it follows that

$$k_m'(x) = 2[k_{m-1}(2x+1) - k_{m-1}(2x-1)] \tag{6.2.4}$$

which shows that formula (6.2.2) is true for $s = 1$ and $m = 2, 3, \ldots$

Let us assume that it is true for some $s \in \mathbb{N}$, $s < m - 1$. Then $s + 1 \leq m - 1$ and

$$k_m^{(s+1)}(x) = (k_m^{(s)}(x))'$$

$$= \left[ 2^{s(s+1)/2} \sum_{i_1=0}^{1} \cdots \sum_{i_s=0}^{1} (-1)^{i_1 + \cdots + i_s} k_{m-s}(2^s x + A_i(s)) \right]'$$

$$= 2^{s(s+1)/2} \sum_{i_1=0}^{1} \cdots \sum_{i_s=0}^{1} (-1)^{i_1 + \cdots + i_s} \cdot 2^s$$

$$\times 2[k_{m-s-1}(2 \cdot 2^s x + 2 \cdot A_i(x) + 1) - k_{m-s-1}(2 \cdot 2^s x + 2 \cdot A_i(s) - 1)]$$

$$= 2^{(s+1)(s+2)/2} \sum_{i_1=0}^{1} \cdots \sum_{i_{s+1}=0}^{1} (-1)^{i_1 + \cdots + i_{s+1}} k_{m-(s+1)}(2^{s+1} x + A_i(s+1))$$

since

$$2 \cdot A_i(s) + (-1)^{i_{s+1}} = 2 \cdot \sum_{j=1}^{s} (-1)^{i_j} \cdot 2^{s-j} + (-1)^{i_{s+1}}$$

$$= \sum_{j=1}^{s+1} (-1)^{i_j} \cdot 2^{s+1-j} = A_i(s+1).$$

If $k(x) \in AC[-1, 1]$, then, according to 7*, $k_m(x) \in C^m(\mathbb{R})$ and the above considerations go for $s = m$ also. The property is proved.

10* If $m \in \mathbb{N}$, $s = 1, 2, \ldots, m - 1$; $v = 0, 1, \ldots, s - 1$, and $p \in \mathbb{Z}$, then

$$k_m^{(s)}(p/2^v) = 0. \tag{6.2.5}$$

If $k(x) \in AC[-1, 1]$, then (6.2.5) is true for $s = m$ also.
Indeed, from (6.2.2) it follows that

$$k_m^{(s)}\left(\frac{p}{2^v}\right) = 2^{s(s+1)/2} \cdot \sum_{i_1=0}^{1} \cdots \sum_{i_s=0}^{1} (-1)^{i_1 + \cdots + i_s} k_{m-s}(2^{s-v} \cdot p + A_i(s)) = 0$$

since $v < s$ and (6.2.3) implies that the number

$$2^{s-v} \cdot p + A_i(s)$$

is odd; hence $2^{s-v} \cdot p + A_i(s) \in \mathbb{R} \setminus (-1, 1)$ from which, according to 2* we have

$$k_{m-s}(2^{s-v} \cdot p + A_i(s)) = 0.$$

If $k(x) \in AC[-1, 1]$, then formula (6.2.2) is true for $s = m$ also, and the above considerations can be made again.

11* The function

$$\overline{k_m}(u) = k_m(u + \tfrac{1}{2}) - \tfrac{1}{2}, \quad u \in [-\tfrac{1}{2}, \tfrac{1}{2}], \quad m \in \mathbb{N} \tag{6.2.6}$$

is odd.

Indeed, for $u \in [-\tfrac{1}{2}, \tfrac{1}{2}]$, we have

(i) $\overline{k_m}(u) = -\tfrac{1}{2} + \int_{2u}^{2u+2} k_{m-1}(t)dt = -\tfrac{1}{2} + \int_{2u}^{1} k_{m-1}(t)dt$

and

(ii) $-u \in [-\tfrac{1}{2}, \tfrac{1}{2}]$

and from (i) and (ii) we get

$$\overline{k_m}(-u) = -\tfrac{1}{2} + \int_{-2u}^{1} k_{m-1}(t)dt$$

$$= -\tfrac{1}{2} + \int_{-2u}^{-1} k_{m-1}(t)dt + \int_{-1}^{1} k_{m-1}(t)dt$$

$$= -\tfrac{1}{2} + \int_{2u}^{1} k_{m-1}(-\tau)d(-\tau) + 1$$

$$= -\left( -\tfrac{1}{2} + \int_{2u}^{1} k_{m-1}(t)dt \right) = -\overline{k_m}(x).$$

Therefore

$$\overline{k_m}(-u) = -\overline{k_m}(u), \quad u \in [-\tfrac{1}{2}, \tfrac{1}{2}], \quad m \in \mathbb{N}. \tag{6.2.7}$$

The property is proved.

From (6.2.7) and $u = 0$ it follows that $\overline{k_m}(0) = 0$, $m \in \mathbb{N}$, i.e.

$$k_m(\tfrac{1}{2}) = \tfrac{1}{2}, \quad m \in \mathbb{N}.$$

This equation along with 8* and property (b) of set $A$ give

12* $k_m(-\tfrac{1}{2}) = k_m(\tfrac{1}{2}) = \tfrac{1}{2}$, $m \in \mathbb{N}$.

Equations (6.2.6) and (6.2.7) imply the identity

$$k_m(u + \tfrac{1}{2}) + k_m(\tfrac{1}{2} - u) = 1, \; u \in [-\tfrac{1}{2}, \tfrac{1}{2}], \; m \in \mathbb{N}. \tag{6.2.7'}$$

and by the substitution $u + \frac{1}{2} = x$ we obtain the following partition of the unit:
13* For each $m \in \mathbb{N}$ it holds that

$$k_m(x) + k_m(1 - x) = 1, \quad x \in [0, 1]. \tag{6.2.8}$$

To facilitate the statements of the following propositions, we remark that the properties 11*–13* express the symmetry of the curve $y = k_m(x)$, $x \in [-1, 0]$ with respect to the point $(-\frac{1}{2}, \frac{1}{2})$ and also the symmetry of the curve $y = k_m(x)$, $0 \leq x \leq 1$ with respect to the point $(\frac{1}{2}, \frac{1}{2})$ for every $m \in \mathbb{N}$. At the same time property 4* expresses the symmetry of the curve $y = k_m(x)$, $0 \leq x \leq 1$ with respect to the ordinate axis.
14* If $m \in \mathbb{N}$, $m \geq 2$, then $k'_m(x) \leq 0$ for every $x \in [0, 1]$.

Indeed, let $x \in [0, 1]$. Then $-1 \leq 2x - 1 \leq 1$, $1 \leq 2x + 1$ and according to 2*, we have $k_{m-1}(2x + 1) = 0$, and by 3*: $k_{m-1}(2x - 1) \geq 0$.

From these relations and from (6.2.4) it follows that $k'_m(x) = -2k_{m-1}(2x - 1) \leq 0$ for every $x \in [0, 1]$.
15* If $m \in fN$, $m \geq 2$, then for each $x \in [0, 1]$ the identity $[k_m(1 - x)]' = -k'_m(x)$ holds. The property follows immediately from (6.2.8) and 7*.
16* If $m \in \mathbb{N}$, $s > 0$, then for each $x \in [0, 1]$ it holds that $k^s_m(x) + k^s_m(1 - x) > 0$. Indeed, according to 3*, the basic quasi-spline is non-negative for each $m \in \mathbb{N}$. and for every $x \in [0, 1]$. According to 13*, for each $x \in [0, 1]$ at least one of the non-negative numbers $k_m(x)$ and $k_m(1 - x)$ is positive and hence $k^s_m(x) + k^s_m(1 - x) > 0$ for each $x \in [0, 1]$.

## 6.3 Convexity of the basic quasi-splines

First, we will remind the reader of some definitions.

A function $f : [a, b] \to \mathbb{R}$ is said to be convex (concave) in the interval $[a, b]$ if for every two points $x$ and $y$ of $[a, b]$ and for every two non-negative numbers $\alpha$ and $\beta$ with $\alpha + \beta = 1$

$$f(\alpha x + \beta y) \leq \alpha \cdot f(x) + \beta \cdot f(y) \tag{6.3.1}$$

or

$$f(\alpha x + \beta y) \geq \alpha \cdot f(x) + \beta \cdot f(y) \tag{6.3.1'}$$

respectively.

If for $\alpha$, $\beta \in (0, 1)$ and $\alpha + \beta = 1$ the equality sign in (6.3.1) (or (6.3.1'), respectively) is attained only for $x = y$, then the function $f$ is said to be strictly convex (concave) in $[a, b]$.

## Lemma 6.3.1

*For every $m \in \mathbb{N}$ the quasi-spline $k_m(x)$, defined in (6.1.2) is a concave function in the interval $\left[-\frac{1}{2}, \frac{1}{2}\right]$.*

The proof will be completed, if we can show that $k_m(x)$ is a concave function in the interval $[0, \frac{1}{2}]$, since the elementary properties 4* and 5* imply the required result.

Let $x, y \in [0, \frac{1}{2}]$ with $x \leq y$. Then for $\alpha \geq 0$, $\beta \geq 0$, $\alpha + \beta = 1$ we have

$$0 \leq x \leq \alpha x + \beta y \leq y \leq 1/2$$

$$-1 \leq 2x - 1 \leq 2(\alpha x + \beta y) - 1 \leq 2y - 1 \leq 0$$

$$1 \leq 2x + 1 \leq 2(\alpha x + \beta y) + 1 \leq 2y + 1$$

and

$$k_m(\alpha x + \beta y) - [\alpha \cdot k_m(x) + \beta \cdot k_m(y)]$$

$$= \int_{2(\alpha x + \beta y) - 1}^{1} k_{m-1}(u)du - \alpha \int_{2x-1}^{1} k_{m-1}(u)du - \beta \int_{2y-1}^{1} k_{m-1}(u)du$$

$$= \beta \int_{2(\alpha x + \beta y) - 1}^{2y-1} k_{m-1}(u)du - \alpha \int_{2x-1}^{2(\alpha x + \beta y) - 1} k_{m-1}(u)du$$

$$\geq \beta \cdot k_{m-1}(2(\alpha + \beta y) - 1) \cdot 2 \cdot (y - \alpha x - \beta y)$$

$$- \alpha \cdot k_{m-1}(2(\alpha x + \beta y) - 1) \cdot 2 (\alpha x + \beta y - x)$$

$$= 2 \cdot k_m(2(\alpha x + \beta y) - 1) \cdot [\beta(y - \alpha x - \beta y) - \alpha(\alpha x + \beta y - x)] = 0$$

since, according to the elementary property 5*, the function $k_{m-1}(x)$, $m \in \mathbb{N}$ is increasing in the segment $[-1, 0]$.

The lemma is proved.

This lemma implies the following result.

**Corollary 6.3.1**   *For every* $x \in [0, \frac{1}{2}]$

$$\delta_1(x) \leq k_{m+1}(x), \quad m \in \mathbb{N}_0$$

$$k_{m+1}(x) \leq \lambda_1(x), \quad m \in \mathbb{N}$$

$$(6.3.2)$$

*where the functions* $\delta_1(x)$ *and* $\lambda_1(x)$ *are defined in* (6.1.8) *and* (6.1.4), *respectively.*

## Proof

Since $m \in \mathbb{N}_0$, then $m + 1 \in \mathbb{N}$ and according to the lemma, the curve $y = k_{m+1}(x)$ is concave in the segment $[0, \frac{1}{2}]$. Furthermore, by the elementary properties 6* and 12* the curve with the equation $y = k_{m+1}(x)$, $m \in \mathbb{N}_0$ and the segment $y = \delta_1(x)$, $0 \leq x \leq \frac{1}{2}$ pass through the points $(0, 1)$ and $(\frac{1}{2}, \frac{1}{2})$. Hence $\delta_1(x) \leq k_{m+1}(x)$, $0 \leq x \leq \frac{1}{2}$ and thus the first of the inequalities, (6.3.2), is proved. The second one is also true, since the curve $y = k_{m+1}(x)$, $0 \leq x \leq \frac{1}{2}$, $m \in \mathbb{N}$ is concave and, according to the elementary property 7*, $k_{m+1}(x) \in C^m(\mathbb{R})$ it has a tangent in each point $x \in [0, \frac{1}{2}]$ and $y = \lambda_1(x), 0 \leq x \leq \frac{1}{2}$ consists of parts of the lines $y = 1$ and $y = (3/2) - 2x$ (see (6.1.4)), which are tangent to this curve for the points $(0, 1)$ and $(\frac{1}{2}, \frac{1}{2})$, respectively.

The first of the inequalities, (6.3.2), is an elementary but is crucial for the following investigation. Let us state it in an explicit form: for each positive integer $m$ and for every $x \in [0, \frac{1}{2}]$

$$k_m(x) \geq 1 - x. \tag{6.3.3}$$

From (6.3.3) and from the partition of the unity (see 13*) it follows that for every positive integer $m$ and for every $x \in [0, \frac{1}{2}]$

$$k_m(1 - x) \leq x. \tag{6.3.4}$$

## Theorem 6.3.1

*For every* $r \in \mathbb{N}_0$ *there exists a function* $b : [0, 1] \to \mathbb{R}_+$ *with the following properties:*
 (i) $b(x)$ *is Riemann-integrable in* $[0, 1]$;
 (ii) $b(x) + b(1 - x) = 1$ *for every* $x \in [0, 1]$;
 (iii) $b(0) = 1$;
 (iv) $x^r \cdot b(x) \geq (1 - x)^r \cdot b(1 - x)$ *for every* $x \in [0, \frac{1}{2}]$ $(0^0 \overset{\text{def}}{=} 1)$.

## Proof

Let $m$ be an arbitrary positive integer, and for each $r \in \mathbb{N}_0$ we write

$$b(x) = \frac{k_m^r(x)}{k_m^r(x) + k_m^r(1 - x)}, \quad x \in [0, 1]. \tag{6.3.5}$$

The function $b(x)$ defined in this way has the properties (i)–(iv). Indeed, from the partition of the unity (see 13*)

$$k_m(x) + k_m(1 - x) = 1, \quad x \in [0, 1]$$

it follows that Definition (6.3.5) of $b(x)$ is satisfied, since $k_m^r(x) + k_m^r(1 - x) \neq 0$ for each $x \in [0, 1]$. It is obvious that $b(x) \geq 0$ for every $x \in [0, 1]$.

From the elementary property 7* it follows that function (6.3.5) is continuous and hence Riemann-integrable in the segment $[0, 1]$.

Since for every $x \in [0, 1]$ we have

$$b(x) + b(1 - x) = \frac{k_m^r(x)}{k_m^r(x) + k_m^r(1 - x)} + \frac{k_m^r(1 - x)}{k_m^r(1 - x) + k_m^r(x)} = 1$$

then the function $b(x)$ has the property (ii). Further,

$$b(0) \overset{(6.3.5)}{=} \frac{k_m^r(0)}{k_m^r(0) + k_m^r(1)} = 1$$

since, as we know, $k_m(0) = 1$ and from the partition of the unity $k_m(0) + k_m(1 - 0) = 1$ we have $k_m(1) = 0$. Hence, $b(x)$ has property (iii).

Finally

$$x^r \cdot b(x) \overset{(6.3.5)}{=} \frac{x^r \cdot k_m^r(x)}{k_m^r(x) + k_m^r(1 - x)} \overset{(6.3.3)}{\geq} \frac{x^r(1 - x)^r}{k_m^r(x) + k_m^r(1 - x)}$$

$$\overset{(6.6.4)}{\geq} \frac{(1 - x)^r \cdot k_m^r(1 - x)}{k_m^r(x) + k_m^r(1 - x)} \overset{(6.3.5)}{=} (1 - x)^r \cdot b(1 - x).$$

The theorem is proved.

In implicit form Theorem 6.3.1 is contained in Kirov [41].

The set of all functions $b(x)$ satisfying the hypothesis of Theorem 6.3.1, is denoted by $A_r$. Each element $b(x)$ of this set is said to be ([42]) a fundamental rational quasi-spline of degree $r$.

Equation (6.3.5) and the elementary property 7* imply that if we choose (independent of $r$) a number $m$ great enough, we can ensure all the necessary derivatives of $b(x)$.

---

**Remark.** In [41] the set $A_r$ is defined for each $r \geq 1$.

---

## 6.4 Some properties of the fundamental rational quasi-splines

Theorem 6.3.1 implies that for each fundamental rational quasi-spline $b \in A_r$, $r \in \mathbb{N}_0$

$$b(\tfrac{1}{2}) = \tfrac{1}{2}, \quad 0 = b(1) \leq b(x) \leq b(0) = 1, \quad x \in [0, 1]. \qquad (6.4.1)$$

### Lemma 6.4.1

*If $r \in \mathbb{N}_0$, $s \in \mathbb{N}_0$ and $s \leq r$, then $A_r \subset A_s$.*

### Proof

For $r = s$ the assertion is trivial.

Let $0 \leq s < r$ and $b \in A_r$. We will show that $b \in A_s$. To this end it is sufficient to prove the inequalities

$$b(x) \geq b(1 - x) \quad \text{for each } x \in [0, \tfrac{1}{2}] \quad \text{when } s = 0 \qquad (6.4.2)$$

and

$$x^s \cdot b(x) \geq (1 - x)^s \cdot b(1 - x) \quad \text{for each } x \in [0, \tfrac{1}{2}] \quad \text{when } 0 < s < r. \qquad (6.4.3)$$

For $x = 0$ inequalities (6.4.2) and (6.4.3) follow from (6.4.1).

If $x \in (0, \tfrac{1}{2}]$, then for every $s \in \mathbb{N}_0$, $s < r$, according to property (ii) of Theorem 6.3.1, we have

$$x^s \cdot b(x) - (1 - x)^s \cdot b(1 - x) \geq (1 - x)^s \cdot b(1 - x) \left[ \left( \frac{1 - x}{x} \right)^{r - s} - 1 \right] \geq 0.$$

The proof is completed.

## Lemma 6.4.2

*If $r \in \mathbb{N}_0$ and $b \in A_r$, then for every $x \in [0, 1]$ and $r \in \mathbb{N}$*

$$x^{r+1}b(x) + (1-x)^{r+1} \cdot b(1-x) \leq \tfrac{1}{2}[x^r \cdot b(x) + (1-x)^r \cdot b(1-x)]. \quad (6.4.4)$$

*For $r = 0$ the corresponding inequality takes the form*

$$x \cdot b(x) + (1-x) \cdot b(1-x) \leq \tfrac{1}{2}. \qquad (6.4.5)$$

## Proof

By the symmetry it is sufficient to prove inequalities (6.4.4) and (6.4.5) for $x \in [0, \tfrac{1}{2}]$. For $x = 0$ these inequalities follow from (6.4.1).

If $x$ belongs to the interval $(0, \tfrac{1}{2}]$, then by the substitutions (though for $r = 0$)

$$\alpha_1 = x, \quad \alpha_2 = 1 - x, \quad \beta_1 = x^r \cdot b(x), \quad \beta_2 = (1-x)^r \cdot b(1-x)$$

from hypothesis (v) of Theorem 6.3.1, we obtain the inequality

$$(\alpha_1 - \alpha_2) \cdot (\beta_1 - \beta_2) \leq 0. \qquad (6.4.6')$$

It is straightforward to see that it is equivalent to the inequality

$$\alpha_1 \cdot \beta_1 + \alpha_2 \cdot \beta_2 \leq \tfrac{1}{2}(\alpha_1 + \alpha_2) \cdot (\beta_1 + \beta_2). \qquad (6.4.7')$$

Since $\alpha_1 + \alpha_2 = 1$, then for the adopted notation, inequality (6.4.7') is equivalent to inequality (6.4.4) for $r \in \mathbb{N}$ and to inequality (6.4.5) for $r = 0$, since according to hypothesis (ii) of Theorem 6.3.1, we have

$$\beta_1 + \beta_2 = b(x) + b(1-x) = 1, \quad x \in [0, 1].$$

The lemma is proved.

Applying Lemma 6.4.1 and Lemma 6.4.2, we obtain the following corollary.

**Corollary 6.4.1**    *Let $r \in \mathbb{N}_0, s \in \mathbb{N}_0, s \leq r + 1$ and $b \in A_r$. Then for each $x \in [0, 1]$[†]*

$$x^s \cdot b(x) + (1-x)^s \cdot b(1-x) \leq 1/2^s \qquad (6.4.6)$$

*with equality for $x = 1/2$.*

---

† In [41] (6.4.6) is proved for $m \geq 2$, $r \geq 1$ and $s \geq 0$.

### Lemma 6.4.3

*Let $r\in\mathbb{N}_0$, $b\in A_r$ and the function $\theta:[0,1]\to\mathbb{R}_+$ is increasing. Then for each $x\in[0,1]$*

$$x^r\cdot b(x)\cdot\theta(x)+(1-x)^r\cdot b(1-x)\cdot\theta(1-x)\leqq\frac{1}{2^{r+1}}[\theta(x)+\theta(1-x)] \quad (6.4.7)$$

*with equality for $x=1/2$.*

### Proof

By the symmetry, it is sufficient to prove (6.4.7) for $x\in[0,1/2]$.
If $0\leq x\leq 1/2$, then $0\leq x\leq 1/2\leq 1-x\leq 1$. Then from the monotonicity of the function $\theta(x)$, it follows that

$$\theta(x)-\theta(1-x)\leqq 0$$

and from hypothesis (v) of Theorem 6.3.1, we obtain the inequality

$$x^r\cdot b(x)-(1-x)^r\cdot b(1-x)\geqq 0.$$

Therefore

$$[x^r\cdot b(x)-(1-x)^r\cdot b(1-x)]\cdot[\theta(x)-\theta(1-x)]\leqq 0.$$

The last inequality, by the equivalence of (6.4.6′) and (6.4.7′), is equivalent to the inequality

$$x^r\cdot b(x)\cdot\theta(x)+(1-x)^r\cdot b(1-x)\cdot\theta(1-x)$$
$$\leqq 1/2[x^r\cdot b(x)+(1-x)^r\cdot b(1-x)]\cdot[\theta(x)+\theta(1-x)]. \quad (6.4.8)$$

Then (6.4.8), (6.4.6) and $\theta(x)\geqq 0$ for every $x\in[0,1]$ imply inequality (6.4.7). From (6.4.1) it follows that for $x=1/2$ we have equality.
The lemma is proved.

**Corollary 6.4.2**    *Let $r\in\mathbb{N}_0$, $b\in A_r$ and the function $\theta:[0,1]\to\mathbb{R}_+$ be increasing and concave. Then for each $x\in[0,1]$*

$$x^r\cdot b(x)\cdot\theta(x)+(1-x)^r\cdot b(1-x)\cdot\theta(1-x)\leq 1/2^r\cdot\theta(\tfrac{1}{2}), \quad (6.4.9)$$

*with equality for $x=1/2$.*

## Proof

From the hypothesis that $\theta(x)$ is concave in the segment $[0, 1]$ it follows that

$$1/2[\theta(x) + \theta(1 - x)] \leqq \theta(\tfrac{1}{2})$$

which by (6.4.7) gives (6.4.9).

---

**Remark.**   The assertions of Lemmas 6.4.2 and 6.4.3 and Corollaries 6.4.1 and 6.4.2 are the basic tools in [42] and they can be found (some of them in weaker form or implicitly) in [41], [43], [44]. Here we give their complete proofs.

---

### Lemma 6.4.4 ([41])

*If $m \in \mathbb{N}, m > 1, r \in \mathbb{N}, b \in A_r,$ then*

$$b^{(\mu)}(0) = b^{(\mu)}(1) = 0 \quad (\mu = 1, 2, \ldots, m - 1) \tag{6.4.10}$$

### Proof

Indeed, from (6.3.5) and from elementary property 7* of the basic quasi-splines it follows that the fundamental rational quasi-spline of degree $r$ belongs to $c^{m-1}\,[0, 1]$.

From (6.3.5) we get

$$b'(x) = \frac{z \cdot k_m^{r-1}(x) \cdot k_m^{r-1}(1 - x)}{[k_m^r(x) + k_m^r(1 - x)]^2} \cdot k_m'(x) \equiv B_r(x) \cdot k_m'(x) \quad x \in [0, 1] \tag{6.4.11}$$

and, hence

$$b^{(\mu)}(x) = \sum_{q=0}^{\mu-1} \binom{\mu - 1}{q} k_m^{(q+1)}(x) \cdot B_r^{(\mu-1-q)}(x), \mu = 1, 2, \ldots, m - 1. \tag{6.4.12}$$

For $\mu = 1$, (6.4.10) follows from (6.4.12) and from the known $k_m'(0) = k_m'(1) = 0$.
For $\mu = 2, \ldots, m - 1$ (6.4.10) follows from (6.4.12) and from property 10* of the basic quasi-splines.
The lemma is proved.

---

**Remark.**   From (6.4.11), $k_m(x) \geqq 0$, and $k_m'(x) \leqq 0$ for every $x \in [0, 1]$ it follows that $b'(x) \leqq 0$ for every $x \in [0, 1]$.

---

## 6.5    Approximation of functions by rational quasi-splines in uniform metrics

The basic results of this section are published in [41], [43] and [44].

Let $n$ be a positive integer, and $\tau_n = (x_0, x_1, \ldots, x_n)$ be an arbitrary mesh†
of the segment $[0, 1]$ with nodes

$$0 = x_0 < x_1 < x_2 < \ldots < x_n = 1.$$

The number $h_i = x_i - x_{i-1}$ $(i = 1, 2, \ldots, n)$, and $|\tau_n| = \max(h_1, \ldots, h_n)$, are
the $i$th step and the diameter of the mesh, correspondingly. Let $\{\tau_n\}$ is the
set of all meshes $\tau_n$ of the segment $[0, 1]$.

Let $r$, $s$ and $v$ be positive integers, $v \leq r$ and assume that the function
$f : [0, 1] \to R$ and that the information

$$T_r = T_r(f, \tau_n) = (f^{(k)}(x_0), \ldots, f^{(k)}(x_n))^r_{k=0} \tag{6.5.1}$$

is given.

Then, for every fundamental rational quasi-spline $b \in A_s$ we introduce ([41],
[42]) a rational quasi-spline of $v$th degree $\Phi : T_r \to R[0, 1]$ by the formula

$$\Phi(x) \equiv \Phi(f; x) \equiv \Phi(b, f_v, \tau_n; x) = b(u) \cdot S_v(f, x_{i-1}; x) + b(1 - u) \cdot S_v(f, x_i; x)$$

$$\tag{6.5.2}$$

if $x \in [x_{i-1}, x_i]$, $i = 1, 2, \ldots, n$, where

$$S_v(f, \alpha; x) = \sum_{k=0}^{v} \frac{f^{(k)}(\alpha)}{k!} (x - \alpha)^k, \quad \alpha \in [0, 1] \tag{6.5.3}$$

and

$$u = (x - x_{i-1})/h_i \quad (i = 1, 2, \ldots, n). \tag{6.5.4}$$

The set of all rational quasi-splines of degree $v$, defined by formula (6.5.2)
is denoted by $\mathscr{L}(A_s, T_v, \{\tau_n\})$. We write $\mathscr{L}_r$ for $\mathscr{L}(A_r, T_r, \{\tau_n\})$.

For $s = 1$ (6.5.2) gives the quasi-splines of order $m$ and of degree $v$, used
in [36].

If $\pi_v$ is the set of all algebraic polynomials of degree $\leq v$ and $f \in \pi_v$, then

$$\Phi(b, f_v, \tau_n; x) = f(x), \quad x \in [0, 1] \tag{6.5.5}$$

since $S_v(f, \alpha; x) = f(x)$ and $b(u) + b(1 - u) = 1$, $u \in [0, 1]$.

If $m = 2, 3, \ldots; v \in \mathbb{N}_0$, $v < m - 1$ and $x_i \in \tau_n$, then

$$\Phi^{(k)}(b, f_v, \tau_n; x_i) = f^{(k)}(x_i) \quad (k = 0, 1, \ldots, v). \tag{6.5.6}$$

Indeed, from (6.5.2) by the Leibnitz formula, taking into account (6.4.1)

---

† The mesh $\tau_n$ has a node more the mesh $X = (x_i, \ldots, x_n)$ used in previous chapters.

and Lemma 6.4.4, we obtain consequently

$$\Phi^{(k)}(b, f_v, \tau_n; x_i) = \sum_{p=0}^{k} \binom{k}{p} \cdot b^{(p)}(1) \cdot S_v^{(k-p)}(f, x_{i-1}; x_i) h_i^{-p}$$

$$+ \sum_{p=0}^{k} \binom{k}{p} \cdot b^{(p)}(0) \cdot S_v^{(k-p)}(f, x_i; x_i) \cdot h_i^{-p} = b(0) \cdot S_v^{(k)}(f, x_i; x_i) = f^{(k)}(x_i)$$

since $b(0) = 1$ and $S_v^{(k)}(f, x_i; x_i) = f^{(k)}(x_i)$.

Therefore, the rational quasi-spline of degree $v$, (6.5.2), under the assumption that we have chosen the order $m$ large enough (for $m > v + 1$), interpolates the function $f(x)$ and its derivatives up to $v$th order inclusively, at the nodes $x_i$ of the mesh $\tau_n$.

The rational quasi-splines (6.5.2) have the property of locality: the properties and the construction of each of them on the subsegment $[x_{i-1}, x_i]$ does not depend on the properties and their construction in other subsegments.

Finally, for large enough $m \in \mathbb{N}$, they have derivatives of at least $v$th order inclusively.

In the special case [41], [45], $s \in \mathbb{N}$, we have

$$b(x) = \frac{\delta_1^s(x)}{\delta_1^s(x) + \delta_1^s(1-x)} = \frac{(1-x)^s}{x^s + (1-x)^s}, \quad x \in [0, 1] \tag{6.5.7}$$

$\Phi(b, f_v, t_n; x)$ is a rational function of $x$ in each of the subsegments $[x_{i-1}, x_i]$ $(i = 1, 2, \ldots, n)$.

Further, it is supposed that the fundamental quasi-spline $b(x)$, defined in (6.3.5) is generated by a function $k(x) \in A$ for $m \in \mathbb{N}$. These conditions, along with the condition $n \in \mathbb{N}$ will not be written explicitly in the following propositions, though they are supposed.

### Theorem 6.5.1 ([41], [43])

*Let $r \in \mathbb{N}$, $s \in \mathbb{N}$, $r \leq s + 1$ and $b \in A_s$. Then*

$$\inf_{\tau_n \in \{\tau_n\}} \sup_{f \in W_\infty^r} \| f(.) - \Phi(b, f_{r-1}, \tau_n; .) \|_C = 1/((2n)^r \cdot r!). \tag{6.5.8}$$

### Proof

Let the hypothesis of the theorem be satisfied, $f \in W_\infty^r$ and $x$ be an arbitrary point of the segment $[0, 1]$, for example $x \in [x_{i-1}, x_i]$ $(i = 1, 2, \ldots, n)$. Then from (6.5.2) and from the partition of the unity $b(u) + b(1 - u) = 1$, $u \in [0, 1]$;

from the Taylor formula with remainder term in integral form (see, e.g. [1], p 25) and from corollary 6.4.1, we get

$$|f(x) - \Phi(b, f_{r-1}, \tau_n; x)|$$

$$= \frac{1}{(r-1)!} \left| b(u) \cdot \int_{x_{i-1}}^{x} (x-t)^{r-1} \cdot f^{(r)}(t)\,dt + b(1-u) \cdot \int_{x_i}^{x} (x-t)^{r-1} \cdot f^{(r)}(t)\,dt \right|$$

$$\leqq \frac{1}{(r-1)!} \left[ b(u) \cdot \int_{x_{i-1}}^{x} (x-t)^{r-1} \cdot 1 \cdot dt + b(1-u) \cdot \int_{x}^{x_i} (t-x)^{r-1} 1 \cdot dt \right]$$

$$= \frac{h_i^r}{r!} \left[ u^r \cdot b(u) + (1-u)^r \cdot b(1-u) \right] \leqq \frac{h_i^r}{2^r \cdot r!}.$$

Hence

$$\sup_{f \in W_\infty^r} \| f(.) - \Phi(b, f_{r-1}, \tau_n; .) \|_C \leqq \frac{|\tau_n|^r}{2^r \cdot r!}. \tag{6.5.9}$$

The equality in (6.5.9) is attained by the function $\bar{f} \in W_\infty^r$ such that

$$\bar{f}^{(r)}(x) = \begin{cases} 0 & \text{for } x \in [0, 1] \setminus [x_{\mu-1}, x_\mu] \\ 1 & \text{for } x \in [x_{\mu-1}, \bar{x}] \\ (-1)^r & \text{for } x \in [\bar{x}, x_\mu] \end{cases}$$

where $\bar{x}$ is the middle point of the segment $[x_{\mu-1}, x_\mu]$, determined by the condition $x_\mu - x_{\mu-1} = h_\mu = |\tau_n|$.

Indeed,

$$\| \bar{f}(.) - \Phi(b, \bar{f}_{r-1}, \tau_n; .) \|_C$$

$$\geqq |\bar{f}(\bar{x}) - \Phi(b, \bar{f}_{r-1}, \tau_n; \bar{x})|$$

$$= \frac{1}{(r-1)!} \left[ b(\tfrac{1}{2}) \cdot \int_{x_{\mu-1}}^{\bar{x}} (\bar{x}-t)^{r-1} \cdot 1 \cdot dt + b(\tfrac{1}{2}) \cdot \int_{\bar{x}}^{x_\mu} (t-\bar{x})^{r-1} (-1)^{2r} \cdot dt \right]$$

$$= \frac{h_\mu^r}{2^r \cdot r!} = \frac{|\tau_n|^r}{2^r \cdot r!}.$$

This relation and (6.5.9) imply

$$\sup_{f \in W^r_\infty} \| f(.) - \Phi(b, f_{r-1}, \tau_m; .) \|_C = \frac{|\tau_n|^r}{2^r \cdot r!}. \tag{6.5.10}$$

Now, from (6.5.10) and from

$$\inf\{ |\tau_n| : \tau_n \in \{\tau_n\} \} = 1/n$$

(6.5.8) follows and the theorem is proved.

If we denote

$$W^r C[0, 1] = W^r C = \{ f(x) : f \in C^r, \| f^{(r)} \|_C \leqq 1 \}$$

then following the proof of Theorem 6.5.1 we will see that

$$\sup_{f \in W^r C} \| f(.) - \Phi(b, f_{r-1}, \tau_n; .) \|_C \leqq \frac{|\tau_n|^r}{2^r \cdot r!}. \tag{6.5.11}$$

Let us consider the function $\tilde{f}(x) \in W^r C[0, 1]$, with

$$\tilde{f}^{(r)}(x) = \begin{cases} 1, & x \in [0, \bar{x} - \varepsilon] \\ 1 + \dfrac{(-1)^r - 1}{2\varepsilon}(x - \bar{x} + \varepsilon), & x \in [\bar{x} - \varepsilon, \bar{x} + \varepsilon] \\ (-1)^r, & x \in [\bar{x} + \varepsilon, 1] \end{cases} \tag{6.5.12}$$

where $\bar{x} = (x_{\mu-1} + x_\mu)/2$ and $0 < \varepsilon < |\tau_n|/4$.

For even $r$ we have $\tilde{f}^{(r)}(x) = \bar{f}^{(r)}(x)$ for $x \in [x_{\mu-1}, x_\mu]$ and

$$\| \tilde{f}(.) - \Phi(b, \tilde{f}_{r=1}, \tau_n; .) \|_C \geqq \tilde{f}(\bar{x}) - \Phi(b, \tilde{f}_{r-1}, \tau_n; \bar{x}) = \frac{|\tau_n|^r}{2^r \cdot r!}$$

and for odd $r$ we have

$$\| \tilde{f}(.) - \Phi(b, \tilde{f}_{r-1}, \tau_n; .) \|_C \geqq \tilde{f}(\bar{x}) - \Phi(b, \tilde{f}_{r-1}, \tau_n; \bar{x}) = \frac{|\tau_n|^r}{2^r \cdot r!} - \frac{\varepsilon^r}{(r+1)!}$$

for every $\varepsilon \in (0, |\tau_n|/4)$.

This, along with (6.5.11) imply the following theorem.

### Theorem 6.5.2 ([11], [43])

*Let $r \in \mathbb{N}$, $s \in \mathbb{N}$, $r \leqq s + 1$ and $b \in A_s$. Then*

$$\inf_{\tau_n \in \{\tau_n\}} \sup_{f \in W^r C} \| f(.) - \Phi(b, f_{r-1}, \tau_n; .) \|_C = 1/((2n)^r \cdot r!). \tag{6.5.13}$$

It is usually said that the modulus of continuity $\omega(t)$ is convex†, if the function $\omega(t)$ is concave over the segment $[0, 1]$.

† Or, 'convex from above'.

## Theorem 6.5.3

*Let $r \in \mathbb{N}$, $s \in \mathbb{N}$, $r \leq s$, $r$ be even, $b \in A_s$, and $\omega(t)$ $(0 \leq t \leq 1)$ be a convex modulus of continuity. Then*

$$\sup_{f \in W^r H^\omega} \| f(.) - \Phi(b, f_r, \tau_n; .) \|_C = \frac{|\tau_n|^r}{2^r \cdot (r-1)!} \int_0^1 (1-z)^{r-1} \cdot \omega\left( \frac{|\tau_n|}{2} z \right) dz \quad (6.5.14)$$

## Proof

For every function $f \in W^r H^\omega$ and for each $x \in [0,1]$, e.g. $x \in [x_{i-1}, x_i]$, $i = 1, 2, \ldots, n$, from (6.5.2) and the hypothesis (ii) of Theorem 6.3.1 and (2.5.1) we obtain ($u$ is determined by $u = (x - x_{i-1})/h_i$) that

$$f(x) - \Phi(b, f_r, \tau_n; x)$$

$$= b(u)[f(x) - S_r(f, x_{i-1}; x)] + b(1-u)[f(x) - S_r(f, x_i; x)]$$

$$= \frac{b(u) \cdot (x - x_{i-1})^r}{(r-1)!} \int_0^1 (1-z)^{r-1} [f^{(r)}(x_{i-1} + z(x - x_{i-1})) - f^{(r)}(x_{i-1})] dz$$

$$+ \frac{b(1-u) \cdot (x - x_i)^r}{(r-1)!} \int_0^1 (1-z)^{r-1} [f^{(r)}(x_i + z(x - x_i)) - f^{(r)}(x_i)] dz.$$

$$(6.5.15)$$

Therefore

$$|f(x) - \Phi(b, f_r, \tau_n; x)|$$

$$\leqq \frac{b(u) \cdot u^r \cdot h_i^r}{(r-1)!} \int_0^1 (1-z)^{r-1} \cdot \omega(z \cdot u h_i) dz$$

$$+ \frac{b(1-u) \cdot (1-u)^r h_i^r}{(r-1)!} \int_0^1 (1-z)^{r-1} \cdot \omega(z(1-u)h_i) dz$$

$$\leqq \frac{|\tau_n|^r}{(r-1)!} \int_0^1 (1-z)^{r-1} [u^r \cdot b(u) \cdot \omega(|\tau_n| \cdot z \cdot u)$$

$$+ (1-u)^r \cdot b(1-u) \cdot \omega(|\tau_n| z \cdot (1-u)) dz. \quad (6.5.16)$$

Since $b \in A_s$ and $r \leqq s$, then by Lemma 6.4.1 $b \in A_r$, and, using Lemma 6.4.3, since $r \in \mathbb{N}$ and the function $\theta(u) = \omega(|\tau_n| \cdot zu)$ is increasing over the segment $[0, 1]$ $(z \in [0, 1])$ then

$$u^r \cdot b(u) \cdot \omega(|\tau_n| \cdot z \cdot u) + (1 - u)^r \cdot b(1 - u) \cdot \omega(|\tau_n| z \cdot (1 - u))$$

$$\leqq \frac{1}{2^{r+1}} \left[ \omega(|\tau_n| z \cdot u) + \omega(|\tau_n| z \cdot (1 - u)) \right]. \quad (6.5.17)$$

Expressions (6.5.16) and (6.5.17) imply

$$|f(x) - \Phi(b, f_r, \tau_n; x)|$$

$$\leqq \frac{|\tau_n|^r}{2^{r+1}(r-1)!} \int_0^1 (1 - z)^{r-1} [\omega(|\tau_n| \cdot z \cdot u) + \omega(|\tau_n| z(1 - u))] dz \quad (6.5.18)$$

Equation (6.5.8) is true for every function $f \in W^r H^\omega$ at each point $x \in [0, 1]$ for an arbitrary (not necessarily convex) modulus of continuity.

If $\omega(t)$ is a convex modulus of continuity, then

$$\frac{1}{2} [\omega(|\tau_n| \cdot z \cdot u) + \omega(|\tau_n| \cdot z \cdot (1 - u))] \leqq \omega\left( \frac{|\tau_n|}{2} z \right). \quad (6.5.19)$$

Now, (6.5.18) and (6.5.19) imply the estimate

$$\sup_{f \in W^r H^\omega} \|f(.) - \Phi(b, f_r, \tau_n; .)\|_C \leqq \frac{|\tau_n|^r}{2^r \cdot (r-1)!} \int_0^1 (1 - z)^{r-1} \cdot \omega\left( \frac{|\tau_n|}{2} \cdot z \right) dz. \quad (6.5.20)$$

Let us consider the function $\bar{f} \in W^r H^\omega$ (with even $r$!) which is an $r$th infinite integral of

$$\bar{f}^{(r)}(x) = \begin{cases} \omega(|x - x_{i-1}|) & \text{for } x \in [x_{i-1}, \bar{x}_i], \\ \omega(|x - x_i|) & \text{for } x \in [\bar{x}_i, x_i], \end{cases} \quad (6.5.21)$$

where $\bar{x}_i = (x_{i-1} + x_i)/2$ $(i = 1, 2, \ldots, n)$.

According to (6.5.15), we have ($r$ is even!)

$$\bar{f}(\bar{x}_\mu) - \Phi(b, \bar{f}_r, \tau_n; \bar{x}_\mu) = \frac{b(\tfrac{1}{2})|\tau_n|^r}{2^r(r-1)!} 2 \int_0^1 (1 - z)^{r-1} \cdot \omega\left( \frac{|\tau_n|}{2} z \right) dz$$

$$= \frac{|\tau_n|^r}{2^r \cdot (r-1)!} \int_0^1 (1 - z)^{r-1} \omega\left( \frac{|\tau_n|}{2} z \right) dz.$$

This and (6.5.20) imply (6.5.14) and thus the theorem is proved.

**Corollary 6.5.1** *Let $r \in \mathbb{N}$, $s \in \mathbb{N}$, $r \leq s$, $r$ be even, $b \in A_s$, and $\omega(t)$ $(0 \leq t \leq 1)$ be a convex modulus of continuity. Then*

$$\inf_{\tau_n \in \{\tau_n\}} \sup_{f \in W^r H^\omega} \| f(.) - \Phi(b, f_r, \tau_n; .) \|_C = \frac{1}{(2n)^r (r-1)!} \int_0^1 (1-z)^{r-1} \omega(z/2n)dz.$$

$$(6.5.22)$$

---

**Remark.** Theorem 6.5.3 and Corollary 6.5.1 are stated and proved here for the first time (weaker variants of these assertions were presented in [41]).

---

**Theorem 6.5.4 ([41])**

*Let $n > 1$, $r \in \mathbb{N}$, $s \in \mathbb{N}$, $r$ be odd $r \leq s$, $b \in A_s$ and $\omega(t) \not\equiv 0$ $(0 \leq t \leq 1)$ be a convex modulus of convexity. Then*

$$\sup_{\omega} \inf_{\tau_n \in \{\tau_n\}} \sup_{f \in W^r H^\omega} \frac{(2n)^r \cdot (r-1)! \, \| f(.) - \Phi(b, f_r, \tau_n; .) \|_C}{\int_0^1 (1-z)^{r-1} \omega(z/2n)dz} = 1. \qquad (6.5.23)$$

The proof of this theorem proceeds in the same way as the proof of Theorem 6.5.3 where the inequality (6.5.20) is obtained. Then (6.5.20) implies for each modulus of continuity $\omega(t)$:

$$\inf_{\tau_n \in \{\tau_n\}} \sup_{f \in W^r H^\omega} \| f(.) - \Phi(b, f_r, \tau_n; .) \|_C \leq \frac{1}{(2n)^r \cdot (r-1)!} \cdot \int_0^1 (1-z)^{r-1} \omega(z/2n)dz.$$

$$(6.5.24)$$

In order to prove the theorem, it remains to show a convex modulus of continuity with respect to a given function, for which the inequality (6.5.24) is satisfied with equality.

Let us consider the convex modulus of continuity

$$\bar{\omega}(t) = \begin{cases} t, & 0 \leq t \leq 1/n \\ 1/n, & 1/n \leq t \leq 1 \end{cases} \qquad (6.5.25)$$

and the function $\hat{f}(t)$ which is an $r$th indefinite integral of the function $\hat{f}^{(r)}(t)$, determined by

$$
\hat{f}^{(r)}(t) = \begin{cases}
\dfrac{1}{2n} - t & \text{for } 0 \le t \le \dfrac{1}{n} \\[2ex]
-\dfrac{3}{2n} + t & \text{for } \dfrac{1}{n} \le t \le \dfrac{2}{n} \\[2ex]
\hat{f}^{(r)}(t) = \hat{f}^{(r)}\left(t - \dfrac{2}{n}\right) & \text{for } \dfrac{2}{n} \le t \le 1
\end{cases}
\tag{6.5.26}
$$

The following properties are obvious:

(a) $\hat{f}^{(r)}(t)$ is continuous;

(b) $\hat{f}^{(r)}(0) = -\hat{f}^{(r)}\left(\dfrac{1}{n}\right) = \dfrac{1}{2n}, \quad \hat{f}^{(r)}\left(\dfrac{1}{2n}\right) = 0$;

(c) $\omega(\hat{f}^{(r)}; t) = \begin{cases}
t, & \text{for } 0 \le t \le \dfrac{1}{n} \\[2ex]
\dfrac{1}{n} & \text{for } \dfrac{1}{n} \le t \le 1
\end{cases}$ .

Therefore $\hat{f} \in W^r H^{\bar{\omega}}$.

Now, for a uniform mesh $\tau_n = \{i/n:\ i = 0, 1, \ldots, n\}$ by (6.5.15) and (a)–(c) (odd $r$) we get

$$
\|\hat{f}(.) - \Phi(b, \hat{f}_r, \tau_n; .)\|_C \ge \left| \hat{f}\left(\dfrac{1}{2n}\right) - \Phi\left(b, \hat{f}_r, \tau_n; \dfrac{1}{2n}\right) \right|
$$

$$
= \left| \dfrac{b(1/2)\cdot(1/2n)^r}{(r-1)!} \int_0^1 (1-z)^{r-1} \left[ \hat{f}^{(r)}\left(\dfrac{z}{2n}\right) - \hat{f}^{(r)}(0) \right] dz \right.
$$

$$
\left. + \dfrac{b(1/2)\cdot(1/2n)^r}{(r-1)!} \int_0^1 (1-z)^{r-1} \left[ \hat{f}^{(r)}\left(\dfrac{1}{n} - \dfrac{z}{2n}\right) - \hat{f}^{(r)}\left(\dfrac{1}{n}\right) \right] dz \right|
$$

$$
= \dfrac{1}{(2n)^r \cdot (r-1)!} \int_0^1 (1-z)^{r-1} \dfrac{z}{2n}\, dz
$$

$$
= \dfrac{1}{(2n)^r \cdot (r-1)!} \int_0^1 (1-z)^{r-1} \bar{\omega}\left(\dfrac{z}{2n}\right) dz.
\tag{6.5.27}
$$

The theorem is proved.

**Corollary 6.5.2** ([41]) *Let $r \in \mathbb{N}$, $s \in \mathbb{N}$, $r \leq s$ and $b \in A_s$. Then for every individual function $f \in C^r$*

$$\| f(.) - \Phi(b, f_r, \tau_n; .) \|_C \leq \frac{|\tau_n|^r}{2^{r-1} \cdot (r-1)!} \cdot \int_0^1 (1-z)^{r-1} \omega\left( f^{(r)}; \frac{|\tau_n|}{2} \cdot z \right) dz.$$

## Proof

By Stechkin's lemma [46], p 78 we construct a convex modulus of continuity $\omega^*(t)$ with the property

$$\omega(f^{(r)}; t) \leq \omega^*(t) \leq 2\omega(f^{(r)}; t). \tag{6.5.28}$$

Then $f \in W^r H^{\omega^*}$ and by (6.5.20)

$$\| f(.) - \Phi(b, f_r, \tau_n; .) \|_C \leq \frac{|\tau_n|^r}{2^r \cdot (r-1)!} \int_0^1 (1-z)^{r-1} \omega^*\left( \frac{|\tau_n|}{2} \cdot z \right) dz$$

$$\overset{(6.5.28)}{\leq} \frac{|\tau_n|^r}{2^{r-1} \cdot (r-1)!} \int_0^1 (1-z)^{r-1} \omega\left( f^{(r)}; \frac{|\tau_n|}{2} \cdot z \right) dz.$$

The proof is completed.

## Theorem 6.5.5 ([41])

*Let $n > 1, r \in \mathbb{N}, s \in \mathbb{N}, r \leq s, b \in A_s$ and $\omega(t) \not\equiv 0$ be a convex modulus of continuity. Then*

$$\sup_\omega \ \inf_{\tau_n \in \{\tau_n\}} \ \sup_{f \in W^r H^\omega} \frac{r! (2n)^r \cdot \| f(.) - \Phi(b, f_r, \tau_n; .) \|_C}{\omega\left( \dfrac{1}{2n(r+1)} \right)} = 1. \tag{6.5.29}$$

## Proof

Let the hypothesis of the theorem be satisfied, and let $r$ be an even number. Then, by Corollary 6.5.1, the equation (6.5.22) holds. For odd $r$ the inequality (6.5.24) is satisfied by Theorem 6.5.4. Hence for every $r \in \mathbb{N}$ the inequality (6.5.24) holds.

According to Problem 75 of Polya and Szego's book ([73], p 53) for functions $p(x)$ and $g(x)$ which are Riemann-integrable in the segment $[a,b]$ with $m \leq g(t) \leq M$, $p(x) \geq 0$ and $\int_a^b p(x)dx > 0$ and for a concave function $\varphi(t)$ in the segment $[m, M]$

$$\frac{\int_a^b p(x) \cdot \varphi[g(x)]dx}{\int_a^b p(x)dx} \leq \varphi\left(\frac{\int_a^b p(x)g(x)dx}{\int_a^b p(x)dx}\right). \tag{6.5.30}$$

We apply inequality (6.5.30) for $p(z) = (1-z)^{r-1} \geq 0$, $g(z) = z/2n$, $z \in [0,1]$, and $\varphi(t) = \omega(t)$. By the hypothesis, the function $\varphi(t) = \omega(t)$ is concave, $p(z)$ and $g(z)$ are, evidently Riemann-integrable on the segment $[0,1]$ and

$$\int_0^1 (1-z)^{r-1}\, dz = 1/r > 0 \tag{6.5.31}$$

$$\int_0^1 (1-z)^{r-1} \cdot \frac{z}{2n} \cdot dz = \frac{1}{2n} \cdot \frac{1}{r(r+1)} \tag{6.5.32}$$

and finally $0 \leq z/2n \leq 1$ for each $z \in [0,1]$ and every $n \in \mathbb{N}$.

Now, (6.5.30)–(6.5.32) imply

$$\int_0^1 (1-z)^{r-1}\omega(z/2n)dz \leq 1/r \cdot \omega(1/2n(r+1)). \tag{6.5.33}$$

Then, (6.5.24) and (6.5.33) imply

$$\inf_{\tau_n \in \{\tau_n\}} \sup_{f \in W^r H^\omega} \| f(.) - \Phi(b, f_r, \tau_n; .)\|_C \leq \frac{1}{(2n)^r \cdot r!} \cdot \omega\left(\frac{1}{2n(r+1)}\right). \tag{6.5.34}$$

In order to complete the proof of the theorem, it is sufficient to show a convex modulus of continuity and a corresponding function, for which equality is attained in (6.5.34).

We use the convex modulus of continuity $\bar{\omega}(t)$, defined in (6.5.25). For odd $t$ we used the function $\hat{f}(t) \in W^r H^{\bar{\omega}}$, defined in (6.5.26). According to (6.5.27) (for a mesh $\tau_n = (0, 1/n, 2/n, \ldots, 1))$ and (6.5.32) we have

$$\| \hat{f}(.) - \Phi(b, \hat{f}_r, \tau_n; .)\|_C \geq |\hat{f}(1/2n) - \Phi(b, \hat{f}_r, \tau_n; 1/2n)|$$

$$= \frac{1}{(2n)^r \cdot (r-1)!} \int_0^1 (1-z)^{r-1} \cdot \frac{z}{2n}\, dz$$

$$= \frac{1}{(2n)^r \cdot r!} \cdot \frac{1}{2n(r+1)} \overset{(6.5.26)}{=} \frac{1}{(2n)^r \cdot r!} \cdot \bar{\omega}\left(\frac{1}{2n(r+1)}\right).$$

Thus the theorem is proved for odd $r$. For even $r$ we use the function $\bar{f}(x)$, which is the $r$th indefinite integral of the function

$$\bar{f}^{(r)}(x) = \begin{cases} x, & 0 \leq x \leq 1/2n \\ 1/n - x, & 1/2n \leq x \leq 1/n \\ \bar{f}^{(r)}(x) = \bar{f}^{(r)}(x - 1/n), & 1/n \leq x \leq 1 \end{cases}$$

and the convex modulus of continuity

$$\bar{\omega}(t) = \begin{cases} t, & 0 \leq t \leq 1/2n \\ 1/2n, & 1/2n \leq t \leq 1 \end{cases}.$$

It is obvious that: (a) $\bar{f} \in W^r H^{\bar{\omega}}$ and (b) $\bar{f}^{(r)}(0) = \bar{f}^{(r)}(1/n) = 0$, $\bar{f}^{(r)}(1/2n) = 1/2n$. Furthermore, according to (6.5.15) for the uniform mesh $\tau_n = (i/n : i = 0, 1, \ldots, n)$ ($r$ is even!) we have

$$\|\bar{f}(.) - \Phi(b, \bar{f}_r, \tau_n; .)\|_C \geq \left| \bar{f}\left(\frac{1}{2n}\right) - \Phi\left(b, \bar{f}_r, \tau_n; \frac{1}{2n}\right) \right|$$

$$= \frac{1}{(2n)^r \cdot (r-1)!} \cdot \int\limits_0^1 (1-z)^{r-1} \frac{z}{2n} \, dz$$

$$= \frac{1}{(2n)^r \cdot r! \, 2n(r+1)}$$

$$= \frac{1}{(2n^r \cdot r!} \, \bar{\omega}\left(\frac{1}{2n \cdot (r+1)}\right).$$

The theorem is proved.

Using Stechkin's lemma, in the same way as we obtained Corollary 6.4.2 from Theorem 6.5.4, from Theorem 6.5.5 we obtain the following corollary.

**Corollary 6.5.3** ([41])   *Let $n > 1$, $r \in \mathbb{N}$, $s \in \mathbb{N}$, $r \leq s$ and $b \in A_s$. Then, for each function $f \in C^r$ ($\tau_n$ is an uniform mesh!)*

$$\|f(.) - \Phi(b, f_r, \tau_n; .)\|_C \leq \frac{2 \cdot \omega(f^{(r)}; 1/2n(r+1))}{(2n)^r \cdot r!}.$$

## Theorem 6.5.6

*Let $r\in\mathbb{N}$, $s\in\mathbb{N}$, $r\leq s$, and $b\in A_s$. Then for every function $f:[0,1]\to\mathbb{R}$ with continuous rth derivative*

$$\|f(.)-\Phi(b,f_r,\tau_n;.)\|_C \leq \frac{|\tau_n|^r}{2^{r+1}\cdot(r+1)!}\cdot \max_{0\leq u\leq 1/2}\int_0^1 (1-z)^{r-1}$$

$$\times [\omega(f^{(r)};|\tau_n|zu)+\omega(f^{(r)};|\tau_n|\cdot z\cdot(1-u))]dz.$$

## Proof

Let the hypothesis of the theorem be satisfied and $x$ be an arbitrary point of the segment $[0,1]$, e.g. $x\in[x_{i-1},x_i]$, $i=1,2,\ldots,n$. Then (6.5.15) implies

$$|f(x)-\Phi(b,f_r,\tau_n;x)|$$

$$\leqq \frac{h_i^r}{(r-1)!}u^r\cdot b(u)\int_0^1 (1-z)^{r-1}\omega(f^{(r)};h;zu)dz$$

$$+\frac{h_i^r}{(r-1)!}(1-u)^r\cdot b(1-u)\cdot\int_0^1 (1-z)^{r-1}\cdot\omega(f^{(r)};h_iz(1-u)dz.$$

From this inequality, from $h_i\leq|\tau_n|$, from the monotonicity of the modulus of continuity of $f^{(r)}(x)$ and by Lemma 6.4.3, we get

$$|f(x)-\Phi(b,f_r,\tau_n;x)| \leqq \frac{|\tau_n|^r}{2^{r+1}\cdot(r+1)!}\int_0^1 (1-z)^{r-1}$$

$$\times [\omega(f^{(r)};|\tau_n|\cdot z\cdot u)+\omega(f^{(r)};|\tau_n|\cdot z\cdot(1-u))]dz. \quad (6.5.35)$$

Denoting

$$\varphi(u)=\omega(f^{(r)};|\tau_n|zu)+\omega(f^{(r)};|\tau_n|z(1-u)), \quad u\in[0,1]$$

it is immediately seen that $\varphi(u) = \varphi(1 - u)$ for each $u \in [0, 1]$. Hence

$$\max_{0 \leq u \leq 1} \varphi(u) = \max_{0 \leq u \leq 1/2} \varphi(u).$$

Using this relation, from (6.5.35) we get

$$\| f(.) - \Phi(b, f_r, \tau_n; .) \|_C \leq \frac{|\tau_n|^r}{2^{r+1} \cdot (r+1)!} \max_{0 \leq u \leq 1/2} \int_0^1 (1 - z)^{r-1}$$

$$\times [\omega(f^{(r)}; |\tau_n|zu) + \omega(f^{(r)}; |\tau_n|z(1 - u))] dz$$

and thus the theorem is proved.

**Corollary 6.5.4**  *Under the hypothesis of Theorem 6.5.6*

$$\| f(.) - \Phi(b, f_r, \tau_n; .) \|_C \leq \frac{|\tau_n|^r}{2^{r+1} \cdot (r-1)!} \int_0^1 (1 - z)^{r-1}$$

$$\times \left[ \omega\left( f^{(r)}; \frac{|\tau_n|}{2} z \right) + \omega(f^{(r)}; |\tau_n| \cdot z) \right] dz \quad (6.5.36)$$

*which implies*

$$\| f(.) - \Phi(b, f_r, \tau_n; .) \|_C \leq \frac{|\tau_n|^r}{2^r \cdot (r-1)!} \cdot \int_0^1 (1 - z)^{r-1}$$

$$\times \min\left\{ \frac{3}{2} \omega\left( f^{(r)}; \frac{|\tau_n|}{2} \cdot z \right), \omega(f^{(r)}; |\tau_n| \cdot z) \right\} dz.$$

Since $\omega(f^{(r)}; |\tau_n|z) \leq 2\omega(f^{(r)}; |\tau_n|/2 \cdot z)$, then the last inequality implies the assertion of Corollary 6.5.2.

**Corollary 6.5.5** (41])  *Let $r \in \mathbb{N}$, $s \in \mathbb{N}$, $r \leq s$ and $b \in A_s$, and $\omega(t)$ be an arbitrary modulus of continuity. Then*

$$\sup_{f \in W^r H^\omega} \| f(.) - \Phi(b, f_r, \tau_n; .) \|_C$$

$$\leq \frac{|\tau_n|^r}{2^{r+1}(r-1)!} \int_0^1 (1-z)^{r-1}\left[\omega\left(\frac{|\tau_n|}{2}\cdot z\right) + \omega(|\tau_n|\cdot z)\right]dz$$

$$\frac{|\tau_n|^r}{2^r(r-1)!} \int_0^1 (1-z)^{r-1} \min\left\{\frac{3}{2}\omega\left(\frac{|\tau_n|}{2}\cdot z\right), \omega(|\tau_n|\cdot z)\right\}dz.$$

So far we have excluded from consideration the case $r = 0$. We now consider this case.

### Theorem 6.5.7

*Let $s \in \mathbb{N}_0$ and $b \in A_s$. Then for each function $f \in C$*

$$\|f(.) - \Phi(b, f_0, \tau_n; .)\|_C \leq \frac{1}{2} \max_{0 \leq u \leq 1/2} [\omega(f; |\tau_n|u) + \omega(f; |\tau_n|(1-u))].$$

$$(6.5.36)$$

Indeed, for each $x \in [0,1]$ and for every function $f \in C$ we have

$$|f(x) - \Phi(b, f_0, \tau_n; x)| = |b(u)\cdot[f(x) - f(x_{i-1})]$$

$$+ b(1-u)[f(x) - f(x_i)]| \leq b(u)\omega(f; h_i u)$$

$$+ b(1-u)\cdot\omega(f; h_i(1-u)) \leq b(u)\omega(f; |\tau_n|\cdot u)$$

$$+ b(1-u)\cdot\omega(f; |\tau_n|(1-u)) \leq \max_{0 \leq u \leq 1} \varphi(u) \qquad (6.5.37)$$

where

$$\varphi(u) = b(u)\cdot\omega(f; |\tau_n|\cdot u) + b(1-u)\omega(f; |\tau_n|(1-u), \quad u \in [0,1].$$

Since $\varphi(0) = \varphi(1) = 0$, $\varphi(1/2) = \omega(f; |\tau_n|1/2)$ and $\varphi(u) = \varphi(1-u)$ for every $u \in [0,1]$, then

$$\max_{0 \leq u \leq 1} \varphi(u) = \max_{0 \leq u \leq 1/2} \varphi(u) \qquad (6.5.38)$$

From the other side, according to Lemma 6.4.3 for $r = 0$ and arbitrary $u \in [0,1]$

$$b(u)\omega(f; |\tau_n|u) + b(1-u)\cdot\omega(f; |\tau_n|(1-u))$$

$$\leq \frac{1}{2}[\omega(f; |\tau_n|\cdot u) + \omega(f; |\tau_n|\cdot(1-u))]. \qquad (6.5.39)$$

Relations (6.5.37)–(6.5.39) imply (6.5.36) and thus the theorem is proved.

**Corollary 6.5.6**   *Let $s \in \mathbb{N}_0$, $b \in A_s$ and $\omega(t)$ $(0 \leq t \leq 1)$ be an arbitrary modulus of continuity. Then*

$$\sup_{f \in H^\omega} \| f(.) - \Phi(b, f_0, \tau_n; .)\|_C \leq \frac{1}{2} \max_{0 \leq u \leq 1/2} \left[ \omega(|\tau_n|u) + \omega(|\tau_n|(1-u)) \right] \quad (6.5.40)$$

*and hence the inequality*

$$\sup_{f \in H^\omega} \| f(.) - \Phi(b, f_0, \tau_n; .)\|_C \leq \min\left\{ \frac{3}{2}\omega\left(\frac{|\tau_n|}{2}\right), \omega(|\tau_n|) \right\}. \quad (6.5.41)$$

*If additionally it is known that $\omega(t)$ is a convex modulus of continuity then*

$$\sup_{f \in H^\omega} \| f(.) - \Phi(b, f_0, \tau_n; .)\|_C = \omega\left(\frac{|\tau_n|}{2}\right). \quad (6.5.42)$$

Indeed, (6.5.40) is implied by (6.5.36) and $f \in H^\omega$, which is equivalent to $\omega(f; \delta) \leqq \omega(\delta)$. Inequality (6.5.41) follows from (6.5.40) and the inequalities

$$\Psi(|\tau_n|) \overset{\text{def}}{=} \max_{0 \leq u \leq 1/2} \left[ \omega(|\tau_n|u) + \omega(|\tau_n|(1-u)) \right] \leqq 2\omega(|\tau_n|)$$

and

$$\Psi(|\tau_n|) \leqq \omega(|\tau_n|/2) + \omega(|\tau_n|) \leqq \omega(|\tau_n|/2) + 2\omega(|\tau_n|/2) = 3\omega(|\tau_n|/2).$$

Inequality (6.5.42) follows from (6.5.40) and from inequality (6.5.19) (with $z = 1$). In (6.5.42) equality is attained by the function

$$\overline{f}(x) = \begin{cases} \omega(|x - x_{i-1}|) & \text{for } x \in [x_{i-1}, \bar{x}_i] \\ \omega(|x - x_i|) & \text{for } x \in [\bar{x}_i, x_i] \end{cases}$$

where $\bar{x}_i = (x_{i-1} + x_i)/2$ $(i = 1, 2, \ldots, n)$. Obviously, $f \in H^\omega$.

**Corollary 6.5.7**   *Let $s \in \mathbb{N}_0$ and $b \in A_s$. Then for every individual function $f \in C$*

$$\| f(.) - \Phi(b, f_0, \tau_n; .)\|_C \leqq \min\left\{ \frac{3}{2}\omega(f; |\tau_n|/2), \omega(f; |\tau_n|) \right\}.$$

## 6.6  Approximation of functions by rational quasi-splines in integral metrics

Here we propose complete proofs of the propositions, announced in [42].

### Theorem 6.6.1 ([42])

*Let $f \in C[0,1]$, $1 \leq p < \infty$, $s \in \mathbb{N}_0$ and $\Phi \in \mathcal{L}(A_s, T_0, \{\tau_n\})$. Then*

$$\|f - \Phi f\|_{L_p[0,1]} \leq \left( \sum_{i=1}^{n} h_i \int_0^1 \omega^p(f; h_i u)\, du \right)^{1/p}. \tag{6.6.1}$$

### Proof

Let $x$ be an arbitrary point of the segment $[0,1]$, e.g. $x \in [x_{i-1}, x_i]$ $(i = 1, 2, \ldots, n)$, and $f \in C$. From the hypothesis $\Phi \in \mathcal{L}(A_s, T_0, \{\tau_n\})$, by (6.5.2) we have

$$\Phi(x) = \Phi(b, f_0, \tau_n; x)$$

$$= b(u) S_0(f, x_{i-1}; x) + b(1-u) S_0(f, x_i; x)$$

$$= b(u) f(x_{i-1}) + b(1-u) f(x_i), \quad u = \frac{x - x_{i-1}}{h_i}, \quad i = 1, 2, \ldots, n. \tag{6.6.2}$$

Then from (6.6.2) and using the partition of the unity

$$(b \in A_s) \quad b(u) + b(1-u) = 1, \quad u \in [0,1]$$

we get

$$f(x) - \Phi(x) = b(u)[f(x) - f(x_{i-1})] + b(1-u)[f(x) - f(x_i)] \tag{6.6.3}$$

and hence

$$|f(x) - \Phi(x)| \leq b(u)\omega(f; x - x_{i-1}) + b(1-u)\omega(f; x_i - x)$$

$$= b(u) \cdot (f; h_i u) + b(1-u) \cdot \omega(f; h_i(1-u)). \tag{6.6.4}$$

From $b \in A_s$, $s \in \mathbb{N}_0$, according to Lemma 6.4.1 it follows that $b \in A_0$. The last relation, together with the increasing of the modulus of continuity $\theta(u) = \omega(f; h_i u)$, $u \in [0,1]$ by Lemma 6.4.3 gives

$$b(u) \cdot \omega(f; h_i u) + b(1-u) \cdot \omega(f; h_i(1-u)) \leq \frac{1}{2}[\omega(f; h_i u) + \omega(f; h_i(1-u))].$$

The last inequality and (6.6.4) give the inequality

$$|f(x) - \Phi(x)| \le \frac{1}{2}[\omega(f;x - x_{i-1}) + \omega(f;x_i - x)] \qquad (6.6.5)$$

for arbitrary $x \in [x_{i-1}, x_i]$ $(i = 1, 2, \ldots, n)$.

Then, using (6.6.5), the triangle inequality and the increasing behaviour of the function $x^p$, $p > 0$, we get

$$\|f - \Phi f\|_{L_p[0,1]}^p = \int_0^1 |f(x) - \Phi(x)|^p \, dx$$

$$= \sum_{i=1}^n \int_{x_{i-1}}^{x_i} |f(x) - \Phi(x)|^p \, dx$$

$$\overset{(6.6.5)}{\le} \sum_{i=1}^n \int_{x_{i-1}}^{x_i} \frac{1}{2^p}[\omega(f;x - x_{i-1}) + \omega(f;x_i - x)]^p \, dx$$

$$\overset{(1.3.1)}{=} \sum_{i=1}^n 2^{-p} \cdot h_i(\|\omega(f;\cdot - x_{i-1}) + \omega(f;x_i - \cdot)\|_{L_p[x_{i-1},x_i]})^p$$

$$\le \sum_{i=1}^n 2^{-p} \cdot h_i(\|\omega(f;\cdot - x_{i-1})\|_{L_p[x_{i-1},x_i]} + \|\omega(f;x_i - \cdot)\|_{L_p[x_{i-1},x_i]})^p$$

$$= \sum_{i=1}^n 2^{-p} \cdot h_i(I_1 + I_2)^p$$

where

$$I_1^p = \frac{1}{h_i} \int_{x_{i-1}}^{x_i} \omega^p(f;x - x_{i-1}) \, dx = \int_0^1 \omega^p(f;h_i u) \, du$$

$$I_2^p = \frac{1}{h_i} \int_{x_{i-1}}^{x_i} \omega^p(f;x_i - x) \, dx = \int_0^1 \omega^p(f;h_i u) \, du = I_1^p.$$

Therefore $I_1 = I_2$, $I_1 + I_2 = 2I_1$ and

$$\|f - \Phi f\|^p_{L_p[0,1]} \leq \sum_{i=1}^{n} h_i \int_0^1 \omega^p(f; h_i u)du.$$

The last inequality implies (6.6.1) and thus the theorem is proved.

Now we will prove a proposition, close to Lemma 2.4.1.

### Lemma 6.6.1

*Let $P$ be an arbitrary non-empty set, $\tau > 0$, $g:P \to [0,\tau]$ and let there exist $x_0 \in P$ such that $g(x_0) = \inf_{x \in P} g(x)$ and the function $f:[0,\tau] \to \mathbb{R}$ be increasing. Then*

$$\inf_{x \in P} f(g(x)) = f(\inf_{x \in P} g(x)). \tag{6.6.6}$$

### Proof

By the hypothesis, for arbitrary element $x \in P$ the value $g(x)$ is a non-negative number. Hence, the set of real numbers $\{g(x): x \in P\}$ is bounded from below (e.g. from 0) and there exists $\inf\{(g(x): x \in P\} = \alpha$. Then, for each $x \in P$

$$g(x) \geq \alpha = g(x_0). \tag{6.6.7}$$

The equation sign in (6.6.7) is given by the hypothesis.

From (6.6.7) and the monotonicity of the function $f$ it follows that for each $x \in P$

$$f(g(x_0)) \leq f(g(x))$$

or, equivalently,

$$\inf_{x \in P} f(g(x)) = f(g(x_0)) = f(\inf_{x \in P} g(x)).$$

Thus the lemma is proved.

**Corollary 6.6.1**　*Let $1 \leq p < \infty$, $s \in \mathbb{N}_0$, $\Phi \in \mathscr{L}(A_s, T_0, \{\tau_n\})$ and $\omega(t)$ $(0 \leq t \leq 1)$ be an arbitrary modulus of continuity. Then*

$$\sup_{f \in H^\omega} \| f - \Phi f \|_{L_p[0,1]} \leq \left( \sum_{i=1}^{n} h_i \int_0^1 \omega^p(h_i u) du \right)^{1/p} \tag{6.6.8}$$

*and*

$$\inf_{\tau_n \in \{\tau_n\}} \sup_{f \in H^\omega} \| f - \Phi f \|_{L_p[0,1]} \leq \left( \int_0^1 \omega^p\left(\frac{u}{n}\right) du \right)^{1/p}. \tag{6.6.9}$$

**Proof**

Inequality (6.6.8) is implied by (6.6.1) and from

$$f \in H^\omega \Leftrightarrow \omega(f; \delta) \leq \omega(\delta).$$

Let

$$P = \{\tau_n\}, \quad g(\tau_n) = \sum_{i=1}^{n} \int_0^{x_i - x_{i-1}} \omega^p(u) du, \quad f(x) = x^{1/p}$$

$x \geq 0$, $p \geq 1$. Since the function $f(x) = x^{1/p}$ is increasing, and the function $\Psi(x) = \int_0^x \omega^p(u) du$ is convex, then by Jensen's inequality we have

$$\inf_{\tau_n \in \{\tau_n\}} g(\tau_n) = \sum_{i=1}^{n} \int_0^{1/n} \omega^p(u) du = g(\tau_n^0) = \int_0^1 \omega^p(u/n) du$$

where $\tau_n^0 = (0, 1/n, 2/n, \ldots, (n-1)/n, 1) \in \{\tau_n\}$. Then, by Lemma 6.6.1,

$$\inf_{\tau_n \in \{\tau_n\}} \sup_{f \in H^\omega} \| f - \Phi f \|_{L_p[0,1]} \leq \inf_{\tau_n \in \{\tau_n\}} \left( \sum_{i=1}^{n} \int_0^{h_i} \omega^p(u) du \right)^{1/p}$$

$$= \left( \inf_{\tau_n \in \{\tau_n\}} \sum_{i=1}^{n} \int_0^{h_i} \omega^p(u) du \right)^{1/p}$$

$$= \left( \int_0^1 \omega^p(u/n)du \right)^{1/p}$$

thus proving inequality (6.6.9).

**Corollary 6.6.2** *Let $\omega(t)$ $(0 \leq t \leq 1)$ be an arbitrary modulus of continuity, $1 \leq p < \infty$, and*

$$T_0(f;\tau_n) = (f(x_0), f(x_1), \ldots, f(x_n))$$

*be an information, formed on a mesh $\tau_n = (x_0, x_1, \ldots, x_n)$ with two fixed nodes $x_0 = 0$ and $x_n = 1$.*

The method of the rational quasi-splines is optimal with respect to the order method for recovery of functions of the class $H^\omega$ on the information $T_0(f;\tau_n)$ in $L_p$ metrics, that is

$$R_{n+1}^2(H^\omega)_{L_p} \leq \inf_{\tau_n \in \{\tau_n\}} \sup_{f \in H^\omega} \|f - \Phi f\|_{L^p} \leq 2 \cdot R_{n+1}^2(H^\omega)_{L_p} \qquad (6.6.10)$$

where $R_{n+1}^2(H^\omega)_{L_p}$ is the error of the optimal method $(2.4.14_2)$.

## Proof

From (6.6.9), from the homogeneity of $\omega(t)$ with respect to positive integer multipliers and from $(2.4.14_2)$ the following chain of inequalities follows:

$$\inf_{\tau_n \in \{\tau_n\}} \sup_{f \in H^\omega} \|f - \Phi f\|_{L_p} \leq \left( \int_0^1 \omega^p(u/n)du \right)^{1/p}$$

$$\leq 2 \cdot \left( \int_0^1 \omega^p(u/2n)du \right)^{1/p} = 2 \cdot R_{n+1}^2(H^\omega)_{L_p}$$

thus proving the right hand inequality of (6.6.10).

The left hand of the inequalities (6.6.10) can be proved in the same way.

Since the function $\bar{f}(x)$, defined in (6.5.43) belongs to the class $H^\omega$, then

$$\sup_{f \in H^\omega} \| f - \Phi f \|_{L_p} \geqq \| \bar{f} - \Phi \bar{f} \|_{L_p}. \tag{6.6.11}$$

Taking into account that $\bar{f}(x_i) = 0$ $(i = 0, 1, 2, \ldots, n)$, we get $\Phi(b, \bar{f}_0, \tau_n; x) = 0$ and hence

$$\| \bar{f} - \Phi \bar{f} \|_{L_p}^p = \int_0^1 |\bar{f}(x) - \Phi(b, \bar{f}_0, \tau_n; x)|^p \, dx$$

$$= \int_0^1 |\bar{f}(x)|^p \, dx$$

$$= \sum_{i=1}^n \left[ \int_{x_{i-1}}^{\bar{x}_i} \omega^p(x - x_{i-1}) dx + \int_{\bar{x}_i}^{x_i} \omega^p(x_i - x) dx \right]$$

$$= 2 \sum_{i=1}^n \int_0^{h_i/2} \omega^p(u) du \geqq 2 \sum_{i=1}^n \int_0^{1/2n} \omega^p(u) du$$

$$= 2n \int_0^{1/2n} \omega^p(u) du$$

$$= \int_0^1 \omega^p\left( \frac{u}{2n} \right) du$$

$$= [R_{n+1}^2(H^\omega)_{L_p}]^p$$

with $\bar{x}_i = (x_{i-1} + x_i)/2$ $(i = 1, \ldots, n)$, i.e.

$$\| \bar{f} - \Phi \bar{f} \|_{L_p} \geqq R_{n+1}^2(H^\omega)_{L_p}. \tag{6.6.12}$$

The inequality follows from the previous chain of relations, using Jensen's inequality.

Then, (6.6.11) and (6.6.12) imply the left-hand-side of inequality (6.6.10), thus proving the corollary.

**Theorem 6.6.2** ([42])

*Let $r \in \mathbb{N}$, $s \in \mathbb{N}$, $r \leq s$, $f \in C^r[0,1]$, and $\Phi \in \mathcal{L}(A_s, T_r, \{\tau_n\})$. Then for each $p \in [1, +\infty)$*

$$\|f - \Phi f\|_{L_p[0,1]} \leq \frac{|\tau_n|^r}{2^r \cdot (r-1)!} \left\{ \sum_{i=1}^{n} h_i \left[ \int_0^1 (1-z)^{r-1} \|\omega(f^{(r)}; h_i z \,\cdot)\|_{L_p[0,1]} dz \right]^p \right\}^{1/p}.$$

$$(6.6.13)$$

**Proof**

Let $x \in [x_{i-1}, x_i]$, $i = 1, 2, \ldots, n$. Then for every $b \in A_s$, using Taylor formula with the remainder in integral form (2.5.1), we get

$$f(x) - \Phi f(x) = f(x) - \Phi(b, f_r, \tau_n; x)$$

$$= \frac{h_i^r}{(r-1)!} \int_0^1 (1-z)^{r-1}$$

$$\times \{ u^r \cdot b(u) [f^{(r)}(x_{i-1} + z(x - x_{i-1})) - f^{(r)}(x_{i-1})]$$

$$+ (1-u)^r \cdot b(1-u) \cdot [f^{(r)}(x_i + z(x - x_i)) - f^{(r)}(x_i)] \} dz$$

$$(6.6.14)$$

with $u = (x - x_{i-1})/h_i$, $h_i = x_i - x_{i-1}$, $i = 1, 2, \ldots, n$.

Then, for every $x \in [x_{i-1}, x_i]$ we obtain the estimate

$$|f(x) - \Phi f(x)| \leq \frac{h_i^r}{(\tau - 1)!} \cdot \int_0^1 (1-z)^{r-1}$$

$$\times [u^r \cdot b(u) \omega(f^{(r)}; h_i z u) + (1-u)^r \cdot b(1-u) \cdot \omega(f^{(r)}; h_i \cdot z \cdot (1-u))] dz.$$

Since the function $\theta(u) = \omega(f^{(r)}; h_i z \cdot u)$, $u \in [0,1]$, is increasing and $b \in A_s$, $r \leq s$ (and, hence, by Lemma 6.4.1, $b \in A_r$) then we can apply Lemma 6.4.3.

Thus we obtain that for each $x \in [x_{i-1}, x_i]$

$$|f(x) - \Phi f(x)| \leq \frac{|\tau_n|^r}{2^{r+1}(r-1)!} \cdot \int_0^1 (1-z)^{r-1}$$

$$\times [\omega(f^{(r)}; z(x - x_{i-1})) + \omega(f^{(r)}; z(x_i - x))]dz. \quad (6.6.15)$$

By (6.6.15), the generalized Minkowski inequality (see, e.g. [6], p 299), the triangle inequality and the increasing behaviour of the function $f(x) = x^p$, $p > 0$, $x \geqq 0$, we obtain

$$\|f - \Phi f\|_{L_p[0,1]} = \left( \int_0^1 |f(x) - \Phi f(x)|^p \, dx \right)^{1/p}$$

$$= \left( \sum_{i=1}^n \int_{x_{i-1}}^{x_i} |f(x) - \Phi f(x)|^p \, dx \right)^{1/p}$$

$$= \left\{ \sum_{i=1}^n h_i [\|f - \Phi f\|_{L_p[x_{i-1}, x_i]}]^p \right\}^{1/p}$$

$$\leqq \left\{ \sum_{i=1}^n h_i \left[ \left\| \frac{|\tau_n|^r}{2^{r+1}(r-1)!} \cdot \int_0^1 (1-z)^{r-1}(\omega(f^{(r)}; z(. - x_{i-1})) \right. \right. \right.$$

$$\left. \left. \left. + \omega(f^{(r)}; z(x_i - .))dz \right\|_{L_p[x_{i-1}, x_i]} \right]^p \right\}^{1/p}$$

$$\leqq \frac{|\tau_n|^r}{2^{r+1}(r-1)!} \left\{ \sum_{i=1}^n h_i \left[ \int_0^1 (1-z)^{r-1} \cdot \|\omega(f^{(r)}; z(. - x_{i-1})) \right. \right.$$

$$\left. \left. + \omega(f^{(r)}; z(x_i - .)) \|_{L_p[x_{i-1}, x_i]} dz \right]^p \right\}^{1/p}$$

$$\leqq \frac{|\tau_n|^r}{2^{r+1}(r-1)!} \left\{ \sum_{i=1}^n h_i \left[ \int_0^1 (1-z)^{r-1}(\|\omega(f^{(r)}; z(. - x_{i-1}))\|_{L_p[i-1, x_i]} \right. \right.$$

$$\left. \left. + \|\omega(f^{(r)}; z(x_i - .))\|_{L_p[x_{i-1}, x_i]})dz \right]^p \right\}^{1/p}$$

$$= \frac{|\tau_n|^r}{2^{r+1}(r-1)!} \left\{ \sum_{i=1}^{n} h_i \left[ \int_0^1 (1-z)^{r-1}(G_1 + G_2)dz \right]^p \right\}^{1/p}$$

where

$$G_1^p = \frac{1}{h_i} \int_{x_{i-1}}^{x_i} \omega^p(f^{(r)}; z(x - x_{i-1}))dx = \int_0^1 \omega^p(f^{(r)}; h_i z u)du$$

$$G_2^p = \frac{1}{h_i} \int_{x_{i-1}}^{x_i} \omega^p(f^{(r)}; z(x_i - x))dx = \int_0^1 \omega^p(f^{(r)}; h_i z u)du = G_1^p$$

or

$$G_1 + G_2 = 2G_1 = 2\|\omega(f^{(r)}; h_i z \,\cdot)\|_{L_p[0,1]}.$$

Therefore

$$\|f - \Phi f\|_{L_p[0,1]} \leqq \frac{|\tau_n|^r}{2^r(r-1)!} \left\{ \sum_{i=1}^{n} h_i \left[ \int_0^1 (1-z)^{r-1}\|\omega(f^{(r)}; h_i z \,\cdot)\|_{L_p} dz \right]^p \right\}^{1/p}.$$

The theorem is proved.

**Corollary 6.6.3**   *Let $r\in\mathbb{N}$, $s\in\mathbb{N}$, $r \leqq s$, $f\in C^r$ and $\Phi\in\mathscr{L}(A_s, T_r, \{\tau_n\})$. Then for every $p\in[1, \infty)$*

$$\|f - \Phi f\|_{L_p[0,1]} \leqq \frac{|\tau_n|^r}{2^r(r-1)!} \int_0^1 (1-z)^{r-1} \cdot \|\omega(f^{(r)}; |\tau_n|z \,\cdot)\|_{L_p[0,1]}dz. \quad (6.6.16)$$

Indeed, from $h_i \leqq |\tau_n|$ and (6.6.13) it follows that

$$\|f - \Phi f\|_{L_p[0,1]} \leqq \frac{|\tau_n|^r}{2^r(r-1)!} \left\{ \left[ \int_0^1 (1-z)^{r-1}\|\omega(f^{(r)}; |\tau_n|z \,\cdot)\|_{L_p[0,1]} dz \right]^p \cdot \sum_{i=1}^{n} h_i \right\}^{1/p}$$

$$= \frac{|\tau_n|^r}{2^r(r-1)!} \cdot \int_0^1 (1-z)^{r-1} \cdot \|\omega(f^{(r)}; |\tau_n|z \,\cdot)\|_{L_p[0,1]}dz.$$

**Corollary 6.6.4** ([42])   *If $r\in\mathbb{N}_0, f\in C^r[0,1]$ and $\Phi\in\mathscr{L}_r$, then for every $p\in[1,\infty)$*

$$\|f-\Phi f\|_{L_p[0,1]} \leqq \frac{|\tau_n|^r}{2^r\cdot r!}\,\|\omega(f^{(r)};|\tau_n|\,.)\|_{L_p[0,1]}. \tag{6.6.17}$$

Indeed, for $r=0$ (6.6.17) is implied by (6.6.1), and for $r\in\mathbb{N}$ of (6.6.16), respectively.

**Corollary 6.6.5**   *Let $r\in\mathbb{N}$, $s\in\mathbb{N}$, $r\leqq s$, $\Phi\in\mathscr{L}(A_s, T_r; \{\tau_n\})$ and $\omega(t)$ be an arbitrary modulus of continuity. Then, for every $p\in[1,\infty)$*

$$\sup_{f\in W^r H^\omega}\|f-\Phi f\|_{L_p[0,1]} \leqq \frac{|\tau_n|^r}{2^r(r-1)!}\int_0^r (1-z)^{r-1}\|\omega(|\tau_n|\cdot z\,.)\|_{L_p[0,1]}\,\mathrm{d}z.$$

Ivanov [47] introduces for every bounded function the local modulus of continuity

$$\omega(f,x;\delta(x))_\infty = \sup\{|\Delta_v f(x)|:|v|\leq\delta(x)\} \tag{6.6.18}$$

where

$$\Delta_v f(x) = \begin{cases} f(x+v)-f(x), & \text{if } x, x+v\in[a,b] \\ 0, & \text{if } x, x+v\notin[a,b] \end{cases} \tag{6.6.19}$$

and $\delta:[a,b]\to\mathbb{R}_+$ is a bounded function.

Evidently, $\omega(f,x;\delta(x))_\infty \geqq 0$ for every $x\in[a,b]$ and if $f:[a,b]\to\mathbb{R}$ is a bounded and measurable function†, and $0\leq\delta_1(x)\leq\delta_2(x)$ (see [48]), then

$$\omega(f,x;\delta_1(x))_\infty \leqq \omega(f,x;\delta_2(x))_\infty. \tag{6.6.20}$$

Following Ivanov [47] we introduce

$$\tau(f,1,\delta)_{\infty,p} = \|\omega(f,.,.\,\delta(.))_\infty\|_{L_p} \tag{6.6.21}$$

for every function $f\in M[a,b]$.

† That is $f\in M[a,b]$.

Let us note some relationships between the recently introduced characteristics. Due to [49], it holds that

$$\omega(f, x; \delta)_\infty \leqq \omega(f, x; 2\delta) \tag{6.6.22}$$

$$\omega(f, x; \delta) \leqq 2 \cdot \omega(f, x; \delta/2)_\infty, \quad \delta = \text{const.} > 0 \tag{6.6.23}$$

$$\tau(f, 1, \delta)_{\infty, p} \leqq \tau(f, 2\delta)_p \leqq 2\tau(f; \delta)_p \tag{6.6.24}$$

$$\tau(f; \delta)_p \leqq 2\tau(f, 1, \delta/2)_{\infty, p} \leqq 2 \cdot \tau(f, 1, \delta)_{\infty, p} \tag{6.6.25}$$

and hence

$$\frac{1}{2}\tau(f, 1, \delta)_{\infty, p} \leqq \tau(f; \delta)_p \leqq 2\tau(f, 1, \delta/2)_{\infty, p} \leqq 2\tau(f, 1, \delta)_{\infty, p}. \tag{6.6.26}$$

### Theorem 6.6.3

*Let $s \in \mathbb{N}_0$, $p \in [1, \infty)$, $f \in M[0, 1]$ and $\Phi \in \mathscr{L}(A_2, T_0; \{\tau_n\})$. Then*

$$\|f - \Phi f\|_{L_p} \leqq \frac{1}{2}\left[ \tau\left(f, 1; \frac{|\tau_n|}{2}\right)_{\infty, p} + \tau(f, 1, |\tau_n|)_{\infty, p} \right]. \tag{6.6.27}$$

### Proof

Let $f \in M[0, 1]$, $\Phi \in \mathscr{L}(A_s, T_0, \{\tau_n\})$ and $x \in [x_{i-1}, x_i]$ Then, equality (6.6.3) holds. Since $|x - x_{i-1}| \leqq h_i u$. $|x - x_i| \leqq h_i(1 - u)$ and $u = (x - x_{i-1})/h_i$ then, by (6.6.18), we have

$$|f(x) - f(x_{i-1})| \leqq \omega(f, x; h_i u)_\infty$$

$$|f(x) - f(x_i)| \leqq \omega(f, x; h_i(1 - u))_\infty.$$

These inequalities, along with (6.6.3) imply the inequality

$$|f(x) - \Phi f(x)| \leqq b(u)\omega(f, x; h_i u)_\infty + b(1 - u)\omega(f, x; h_i(1 - u))_\infty. \tag{6.6.28}$$

Since the function $\theta(u) = \omega(f, x; h_i u)_\infty$ is non-negative and increasing, and

$b \in A_s$, $0 \le s$, then $b \in A_0$ and by Lemma 6.4.3 (for $r = 0$) we have

$$b(u)\omega(f, x, h_i u)_\infty + b(1 - u)\omega(f, x, h_i(1 - u))_\infty$$

$$\le \frac{1}{2}\left[\omega(f, x, h_i u)_\infty + \omega(f, x, h_i(1 - u))_\infty\right]. \quad (6.6.29)$$

From (6.6.28) and (6.6.29) we obtain the inequality

$$|f(x) - \Phi f(x)| \le \frac{1}{2}\left[\omega(f, x, x - x_{i-1})_\infty + \omega(f, x, x_i - x))_\infty\right] \quad (6.6.30)$$

valid for $x_{i-1} \le x \le x_i$.

On the other hand, if $x_{i-1} \le x \le (x_{i-1} + x_i)/2$, then

$$\omega(f, x, x - x_{i-1})_\infty = \sup\{|f(x) - f(x_{i-1})| : |x - x_{i-1}| \le x - x_{i-1}\}$$

$$\le \sup\{|f(x) - f(x_{i-1})| : |x - x_{i-1}| \le h_i/2\}$$

$$= \omega(f, x, h_i/2)_\infty$$

and, similarly

$$\omega(f, x, x_i - x)_\infty \le \omega(f, x, h_i)_\infty.$$

Therefore, if $x_{i-1} \le x \le (x_{i-1} + x_i)/2$, then

$$\omega(f, x, x - x_{i-1})_\infty + \omega(f, x, x_i - x)_\infty \le \omega(f, x, h_i/2)_\infty + \omega(f, x, h_i)_\infty. \quad (6.6.31)$$

If, additionally $(x_{i-1} + x_i)/2 \le x \le x_i$, then

$$\omega(f, x, x - x_{i-1})_\infty \le \omega(f, x; h_i)_\infty$$

and

$$\omega(f, x, x_i - x)_\infty \le \omega(f, x, h_i/2)_\infty$$

and (6.6.31) follows again, i.e. inequality (6.6.31) is true for each $x \in [x_{i-1}, x_i]$. Then, (6.6.30) (6.6.31), $h_i \le |\tau_n|$ and (6.6.20) imply the validity of the inequality

$$|f(x) - \Phi f(x)| \le \frac{1}{2}\left[\omega(f, x, |\tau_n|/2)_\infty + \omega(f, x, |\tau_n|)_\infty\right] \quad (6.6.32)$$

for every $x \in [0, 1]$.

Using the triangle inequality, from (6.6.32) by (6.6.21) we get

$$\|f - \Phi f\|_{L_p[0,1]} \leq \frac{1}{2}\left[\tau\left(f, 1; \frac{|\tau_n|}{2}\right)_{\infty,p} + \tau(f, 1; |\tau_n|)_{\infty,p}\right]$$

and thus the proof is completed.

**Corollary 6.6.6** ([42])   *If* $p \in [1, \infty)$, $s \in \mathbb{N}_0$, $f \in R[0.1]$ *and* $\Phi \in \mathscr{L}(A_s, T_0, \{\tau_n\})$, *then*

$$\|f - \Phi f\|_{L_p[0,1]} \leq \frac{1}{2}[\tau(f; |\tau_n|)_p + \tau(f; 2|\tau_n|)_p] \leq \frac{3}{2}\tau(f; |\tau_n|)_p. \quad (6.6.33)$$

Indeed, the first of inequalities (6.6.33) follows from $R[0,1] \subset M[0,1]$, (6.6.27) and (6.6.24). The second of inequalities (6.6.33) follows from the first of them and from the property that the averaged modulus of smoothness $\tau(f; \delta)_p$ (see, for example [2]):

$$\tau(f; n\delta)_p \leq n \cdot \tau(f; \delta)_p, \quad n \in \mathbb{N}.$$

For the following considerations we need a local modulus of continuity for the bounded function $f : [0, 1] \to \mathbb{R}$ at a point $x \in [0, 1]$:

$$\omega^*(f, x; \delta(x)) = \sup\left\{|f(t + h) - f(t)| : t, t + h \in \left[x - \frac{\delta(x)}{2}, x + \frac{\delta(x)}{2}\right] \cap [0, 1]\right\}$$

$$(6.6.34)$$

where $\delta : [0, 1] \to [0, 1]$.
   The inequality

$$\omega^*(f, x; \delta(x)) \geq 0 \qquad\qquad (6.6.35)$$

hold for every $x \in [0, 1]$.
   Then, from $0 \leq \delta_1(x) \leq \delta_2(x) \leq \delta \leq 1$, $x \in [0, 1]$, $\delta = \text{const.} > 0$ it follows that

$$0 \leq \omega^*(f, x; \delta_1(x)) \leq \omega^*(f, x; \delta_2(x)) \leq \omega(f, x, \delta). \qquad (6.6.36)$$

Inequality (6.6.35) follows at once from (6.6.34), and inequalities (6.6.36) by (6.6.34) and (1.3.4) are equivalent to the inequalities

$$\sup\{|f(t + h) - f(t)| : t, t + h \in \left[x - \frac{\delta_1(x)}{2}, x + \frac{\delta_1(x)}{2}\right] \cap [0, 1]\}$$

$$\leq \sup\{|f(t+h)-f(t)|:t,t+h\in\left[x-\frac{\delta_2(x)}{2},x+\frac{\delta_2(x)}{2}\right]\cap[0,1]\}$$

$$\leq \sup\{|f(t+h)-f(t)|:t,t+h\in\left[x-\frac{\delta}{2},x+\frac{\delta}{2}\right]\cap[0,1]\}.$$

These hold, since the supremum of a set of real numbers does not decrease by extending the set.

As previously stated, by $L_p^r[0,1]=L_p^r(r\in\mathbb{N},\ 1\leq p<+\infty)$, we denote the set of all functions $f:[0,1]\to\mathbb{R}$ with absolutely continuous $(r-1)$th derivative, such that the $r$th derivative belongs to $L_p[0,1]$. For each function $f$ of $L_p^r$ we suppose it to be given its value $f(x)$ in each point $x\in[0,1]$, thus not identifying it with its equivalence class.

## Theorem 6.6.4

*Let $r\in\mathbb{N}$, $s\in\mathbb{N}$, $r\leq s$, $1\leq p<\infty$, $f\in L_p^r$ and $\Phi\in\mathscr{L}(A_s,T_r,\{\tau_n\})$. Then*

$$\|f-\Phi f\|_{L_p[0,1]}\leq\frac{|\tau_n|^r}{2^r\cdot r!}\cdot\frac{1}{2}[\tau(f^{(r)};|\tau_n|)_p+\tau(f^{(r)};2|\tau_n|)_p] \qquad (6.6.37)$$

*where $\tau(f;\delta)_p$ is the averaged modulus of smoothness, defined in (1.3.5).*

## Proof

Let $x\in[x_{i-1},x_i]$ $(i=1,2,\ldots,n)$, $r\in\mathbb{N}$, $s\in\mathbb{N}$, $r\leq s$, $1\leq p<\infty$, $f\in L_p^r$ and $\Phi\in\mathscr{L}(A_s,T_r,\{\tau_n\})$. Then, as in (6.6.14) (Taylor formula (2.5.1) remains valid, see [1], p 25), we obtain the estimate

$$|f(x)-\Phi f(x)|\leq\frac{h_i^r}{(\tau-1)!}\int_0^1(1-z)^{r-1}\{u^r\cdot b(u)\cdot|f^{(r)}(x_{i-1}+z(x-x_{i-1}))$$

$$-f^{(r)}(x_i)|+(1-u)^r\cdot b(1-u)\cdot|f^{(r)}(x_i+z(x-x_i))-f^{(r)}(x_i)|\}dz \qquad (6.6.38)$$

where $b\in A_s$, $u=(x-x_{i-1})/h_i$, $h_i=x_i-x_{i-1}$, $i=1,2,\ldots,n$.
Since

$$|x-x_{i-1}|\leq h_iu,\quad |x-x_{i-1}-z(x-x_{i-1})|\leq h_iu(1-z)\leq h_iu$$

and

$$|x - x_i| \leq h_i(1 - u), \quad |x - x_i - z(x - x_i)| \leq h_i(1 - u)(1 - z) \leq h_i(1 - u)$$

then, according to (6.6.34)

$$|f^{(r)}(x_{i-1}) + z(x - x_{i-1})) - f^{(r)}(x_{i-1})| \leq \omega^*(f^{(r)}, x; 2h_i u)$$

$$|f^{(r)}(x_i) + z(x - x_i)) - f^{(r)}(x_i)| \leq \omega^*(f^{(r)}, x; 2h_i(1 - u)).$$

These estimates, along with (6.6.38) give the estimate

$$|f(x) - \Phi f(x)| \leq \frac{|\tau_n|^r}{\tau!} \cdot [u^r \cdot b(u) \cdot \omega^*(f^{(r)}, x; 2h_i u)$$

$$+ (1 - u)^r \cdot b(1 - u) \cdot \omega^*(f^{(r)}, x; 2h_i(1 - u))]$$

whence, keeping in mind (6.6.35) and (6.6.36), by Lemma 6.4.3, we get

$$|f(x) - \Phi f(x)| \leq \frac{|\tau_n|^r}{2^{r+1} \cdot r!} \cdot [\omega^*(f^{(r)}, x; 2(x - x_{i-1})) + \omega^*(f^{(r)}, x; 2(x_i - x))].$$

$$(6.6.39)$$

Proceeding as in the proof of (6.6.31), we obtain that for each $x \in [x_{i-1}, x_i]$, $i = 1, 2, \ldots, n$

$$\omega^*(f^{(r)}, x; 2(x - x_{i-1}) + \omega^*(f^{(r)}, x; 2(x_i - x) \leq \omega^*(f^{(r)}, x; h_i) + \omega^*(f^{(r)}, x; 2h_i).$$

Using this result, (6.6.39) and once again (6.6.36), we obtain for each $x \in [0, 1]$

$$|f(x) - \Phi f(x)| \leq \frac{|\tau_n|^r}{2^{r+1} \cdot r!} [\omega^*(f^{(r)}, x; |\tau_n|) + \omega^*(f^{(r)}, x; 2|\tau_n|)].$$

Further, taking the $L_p$ norm, we obtain

$$\|f - \Phi f\|_{L_p[0,1]} \leq \frac{|\tau_n|^r}{2^{r+1} r!} \cdot [\tau(f^{(r)}; |\tau_n|)_p + \tau(f^{(r)}; 2|\tau_n|)_p] \qquad (6.6.40)$$

since, according to (6.6.34) and (1.3.4) with $\delta(x) = \delta = \text{const.} > 0$ we get

$$\omega^*(f, x; \delta) = \sup\{|f(t + h) - f(t)| : t, t + h \in [x - (\delta/2), x + (\delta/2)] \cap [0, 1]\}$$

$$= \omega(f, x; \delta)$$

and hence, by (1.3.5),

$$\|\omega^*(f, x; \delta)\|_{L_p[0,1]} = \tau(f; \delta)_p.$$

The theorem is proved.

**Corollary 6.6.7** *If $1 \leq p \leq \infty$, $r \in \mathbb{N}$, $s \in \mathbb{N}$, $r \leq s$, $f \in L_p^r$ and $\Phi \in \mathcal{L}(A_s, T_r, \{\tau_n\})$, then*

$$\|f - \Phi f\|_{L_p[0,1]} \leq \frac{|\tau_n|^r}{2^r \cdot r!} \cdot \frac{3}{2} \tau(f^{(r)}; |\tau_n|)_p.$$

Finally, we give an application of the results obtained to the approximate evaluation of definite integrals.

Using the information

$$T_r(f, \tau_n) = (f^{(r)}(x_0), f^{(k)}(x_1), \ldots, f^{(k)}(x_n))_{k=0}^r, \quad n \in \mathbb{N}, \quad r \in \mathbb{N}_0$$

we construct the quadrature formula

$$\int_0^1 f(x)\mathrm{d}x = Q_r(f, \tau_n) + R_{nr}(f) \tag{6.6.41}$$

with

$$Q_r(f, \tau_n) = \sum_{i=0}^n \sum_{k=0}^r B_{ik} \cdot f^{(k)}(x_i) \tag{6.6.42}$$

$$B_{ik} = \frac{h_{i+1}^{k+1} + (-1)^k \cdot h_i^{k+1}}{k!} \cdot \int_0^1 b(u) u^k \, \mathrm{d}u \tag{6.6.43}$$

$$i = 0, 1, 2, \ldots, n, \quad k = 0, 1, \ldots, r, \quad h_0 = h_{n+1} = 0, \quad b \in A_s, \quad r \leq s.$$

First, let us calculate

$$\int_0^1 \Phi(b, f_r, \tau_n; x)\mathrm{d}x = \sum_{i=1}^n \int_{x_{i-1}}^{x_i} \Phi(b, f_2, \tau_n; x)\mathrm{d}x$$

$$= \sum_{i=1}^{n} \sum_{k=0}^{r} \frac{1}{k!} \left[ f^{(k)}(x_{i-1}) \cdot \int_{x_{i-1}}^{x_i} b(u) \cdot (x - x_{i-1})^k \, dx \right.$$

$$+ f^{(k)}(x_i) \cdot \int_{x_{i-1}}^{x_i} b(1-u) \cdot (x - x_i)^k \, dx.$$

Since $u = (x - x_{i-1})/h_i$ and

$$\int_{x_{i-1}}^{x_i} b(u) \cdot (x - x_{i-1})^k \, dx = h_i^{k+1} \cdot \int_0^1 b(u) \cdot u^k \, du$$

$$\int_{x_{i-1}}^{x_i} b(1-u) \cdot (x - x_i)^k \, dx = (-1)^k \cdot h_i^{k+1} \cdot \int_0^1 b(v) \cdot v^k \, dv$$

then

$$\int_0^1 \Phi(b, f_r, \tau_n; x) \, dx$$

$$= \sum_{i=1}^{n} \sum_{k=0}^{r} \frac{1}{k!} \left[ h_i^{k+1} \cdot f^{(k)}(x_{i-1}) + (-1)^k \cdot h_i^{k+1} \cdot f^{(k)}(x_i) \right] \cdot \int_0^1 b(u) \cdot u^k \, du$$

$$= \sum_{i=0}^{n} \sum_{k=0}^{r} \frac{h_{i+1}^{k+1} + (-1)^k \cdot h_i^{k+1}}{k!} \cdot \int_0^1 b(u) \cdot u^k \, du \cdot f^{(k)}(x_i) \equiv Q_r(f, \tau_n)$$

that is

$$Q_r(f, \tau_n) = \int_0^1 \Phi(b, f_r, \tau_n; x) \, dx. \tag{6.6.44}$$

Next

$$|R_{nr}(f)| = \left| \int_0^1 f(x) \, dx - Q_r(f, \tau_n) \right| \overset{(6.6.44)}{=} \left| \int_0^1 [f(x) - \Phi(b, f_r, \tau_n; x)] \, dx \right|$$

$$\leq \int_0^1 |f(x) - \Phi(b, f_r, \tau_n; x)|\,dx = \|f - \Phi f\|_{L_1[0,1]}$$

that is

$$|R_{nr}(f)| \leq \|f - \Phi f\|_{L_1[0,1]} \leq \|f - \Phi f\|_{C[0,1]}. \tag{6.6.45}$$

Then (6.6.45) and Corollary 6.6.4 (with $r = 0$ and $p = 1$) imply the assertion of the following corollary.

**Corollary 6.6.8** ([42])   *If* $f \in C[0,1]$, *then*

$$|R_{no}(f)| \leq \int_0^1 \omega(f; |\tau_n|u)\,du.$$

**Corollary 6.6.9** ([42])   *If* $f \in C^r[0,1]$, $r \in \mathbb{N}$, *then*

$$|R_{nr}(f)| \leq \frac{|\tau_n|^r}{2^r \cdot (r-1)!} \cdot \int_0^1 \int_0^1 (1-z)^{r-1} \cdot \omega(f^{(r)}; |\tau_n| \cdot z \cdot u)\,du\,dz. \tag{6.6.46}$$

Indeed, inequality (6.6.46) follows from (6.6.45) and Corollary 6.5.3 for $p = 1$.

**Corollary 6.6.10** ([42])   *If* $r \in \mathbb{N}_0$, $f \in W^r R[0,1]$, *then*

$$|R_{nr}(f)| \leq \frac{|\tau_n|^r}{2^r \cdot r!} \cdot \tau(f^{(r)}; |\tau_n|)_{L_1[0,1]}. \tag{6.6.47}$$

Indeed, inequality (6.6.47) follows from (6.6.45) and (6.6.35) with $r = 0$ and $p = 1$, (6.6.45) and from Corollary 6.6.7 with $r \in \mathbb{N}$ and $p = 1$ provided we note that $f \in R[0,1]$ implies $f \in L[0,1]$.

In the special case when as a fundamental rational quasi-spline it uses function (6.5.7), then by formula (6.5.2) we obtain the rational spline of degree $r$ as follows:

$$\Phi(b, f_r, \tau_n; x)$$

$$= \sum_{k=0}^{r} \frac{f^{(k)}(x_{i-1}) \cdot (x - x_{i-1})^k \cdot (x_i - x)^r + f^{(k)}(x_i)(x - x_i)^k \cdot (x - x_{i-1})^r}{k![(x_i - x)^r + (x - x_{i-1})^r]} \qquad (6.6.48)$$

for $x_{i-1} \leqq x \leqq x_i$, $i = 1, 2, \ldots, n$.

Since the rational spline (6.6.48) is a special case of (6.5.2), there still remain results to be obtained in sections 6.5 and 6.6.

# 7

---

# Applications

In this chapter the theory of the atomar function [50], [51] is developed in relation to the basic quasi-splines, since, in the author's opinion, using this approach, its approximation properties and its relation to the approximation of functions with quasi-splines are much clearer.

## 7.1 Representation of the derivatives of the basic quasi-splines

First we will obtain other formulas for calculating an arbitrary derivative of a basic quasi-spline. Using them, it is possible to generalize the identity, expressing the partition of the unity. (See 13*.)

**Lemma 7.1.1** ([52])

*If* $m = 2, 3, \ldots$ ; $s = 1, 2, \ldots, m - 1$ *then*

$$k_m^{(s)}(x) = 2^{s(s+1)/2} \sum_{i=0}^{2^s - 1} (-1)^{\sigma(i)} k_{m-s}(2^s(x+1) - 1 - 2i) \qquad (7.1.1)$$

*with* $x \in \mathbb{R}$;

$$\sigma(i) = \sum_{j=1}^{s} i_j \qquad (7.1.2)$$

*is the sum of the double digits* $i_j$ *(i.e.* $i_j = 0$, *or* $i_j = 1$ *for* $j = 1, 2, \ldots, s$) *of the s-digital binary representation of the number*

$$i = \sum_{j=1}^{s} i_j 2^{s-j} \qquad (7.1.3)$$

*and $k_m(x)$ is the basic quasi-spline of order $m$ (see (6.1.1) and (6.1.2)).*

The proof of the lemma follows from (6.2.2), (7.1.1) and from the following facts:

($a$) the number of the terms in the left-hand-side of (6.2.2) is $2^s$, exactly so, as in the right-hand-side of (7.1.1);

($b$) to each term in (6.2.2), determined by the sequence $i = (i_1, i_2, \ldots, i_s)$, of the binary digits $i_1, i_2, \ldots, i_s$, we put into correspondence those terms in (7.1.1), which are determined by the number (7.1.3);

($c$) the corresponding terms in (6.2.2) and (7.1.1) are equal.

Indeed, from (6.2.2) and (7.1.3) the chain of equalities

$$A_i(s) = \sum_{j=1}^{s} (-1)^{i_j} \cdot 2^{s-j} = \sum_{j=1}^{s} 2^{s-j} - \sum_{j=1}^{s} [1 - (-1)^{i_j}] 2^{s-j}$$

$$= 2^s - 1 - 2 \cdot \sum_{j=1}^{s} i_j \cdot 2^{s-j} = 2^s - 1 - 2i$$

follows, since $1 - (-1)^{i_j} = 2i_j$ for every binary digit $i_j$.

This relation and (7.1.2) imply

$$(-1)^{i_j + \ldots + i_s} k_{m-s}(2^s x + A_i(s)) = (-1)^{\sigma(i)} k_{m-s}(2^s(x+1) - 1 - 2i)$$

and thus the proof is completed.

Further, along with the notation $\sigma(.)$ for the sum of the binary digits of an integer, we use the notation $[y]$ and $\{y\}$ for the integer and for the fractional part of the number $y$ respectively, i.e. $[y]$ is the greater integer not exceeding $y$ and $\{y\} = y - [y]$.

**Theorem 7.1.1 ([52])**

*If $m = 2, 3, \ldots$; $s = 1, 2, \ldots, m-1$ and $x \in [-1, 1]$, then*

$$k_m^{(s)}(x) = (-1)^{\sigma([2^{s-1}(x+1)]} \cdot 2^{s(s+1)/2} \cdot k_{m-s}(-1 + 2\{2^{s-1} x\}). \qquad (7.1.4)$$

## Proof

For $x = 1$ equality (7.1.4) can be checked immediately using the elementary properties 2* and 10* of the basic quasi-splines. It remains to be proven for $x \in [-1, 1)$. In this case the following auxiliary assertion is true.

In the set

$$\{0, 1, 2, \ldots, 2^s - 1\} \tag{7.1.5}$$

there is only one number $i$ such that the equality

$$2^s(x + 1) - 1 - 2i = -1 + 2\{2^{s-1}x\} \tag{7.1.6}$$

holds, and that

$$-1 \leq 2^s(x + 1) - 1 - 2i < 1. \tag{7.1.7}$$

The uniqueness of the number $i$ with the desired properties follows from (7.1.7). Indeed, if we assume the existence of two integers $i$ and $j$ from the set (7.1.5), satisfying inequalities (7.1.7), the difference $j - i$ would be an integer satisfying the inequalities

$$-1 < j - i < 1$$

which is possible only for $j - i = 0$, i.e. for $j = i$.

The existence will be shown, together with a method of construction. We will define

$$i \stackrel{\text{def}}{=} [2^{s-1}(x + 1)]. \tag{7.1.8}$$

Indeed, the number (7.1.8) is an integer by definition. From $x \in [-1, 1]$ it follows that

$$i = [2^{s-1}(x + 1)] \leq 2^{s-1}(x + 1) < 2^s \Rightarrow i \leq 2^s - 1 \tag{7.1.9}$$

and

$$i = [2^{s-1}(x + 1)] > 2^{s-1}(x + 1) - 1 \geq -1 \Rightarrow i \geq 0. \tag{7.1.10}$$

From (7.1.9) and (7.1.10) it follows that the number (7.1.8) belongs to the set (7.1.5).

Using (7.1.8), we obtain

$$2^s(x+1) - 1 - 2i = 2^s(x+1) - 1 - 2\cdot[2^{s-1}(x+1)]$$
$$= 2(2^{s-1}(x+1) - [2^{s-1}(x+1)]) - 1$$
$$= -1 + 2\cdot\{2^{s-1}(x+1)\}$$
$$= -1 + 2\cdot\{2^{s-1}x + 2^{s-1}\}$$
$$= -1 + 2\cdot\{2^{s-1}x\}$$

that is, equality (7.1.6).

Inequalities (7.1.7) are equivalent to the inequalities

$$0 \le 2^{s-1}(x+1) - [2^{s-1}(x+1)] < 1$$

which are obviously satisfied, since the fractional part of every real number is in the interval $[0, 1)$.

The auxiliary proposition is proved.

From the auxiliary proposition it follows that for every integer $j$ from the set (7.1.5), and different from (7.1.8), we have $2^s(x+1) - 1 - 2j \notin [-1, 1)$ for each $x \in [-1, 1)$ and hence

$$k_{m-s}(2^s(x+1) - 1 - 2j) = 0.$$

This, along with (7.1.1) and (7.1.6) implies (7.1.4) and thus the theorem is proved.

**Corollary 7.1.1**  *If $m = 2, 3, 4, \ldots$; $s = 1, 2, \ldots, m - 1$; $x \in [-1, 1 - (1/2^{s-1})]$, then*

$$k_m^{(s)}(x) = (-1)^{\sigma([2^{s-1}(x+1)]) - \sigma([2^{s-1}(x+1)+1])} \cdot k_m^{(s)}\left(x + \frac{1}{2^{s-1}}\right). \qquad (7.1.11)$$

## Proof

If $x \in [-1, x - (1/2^{s-1})]$, then $x$ and $x + (1/2^{s-1})$ belong to the segment $[-1, 1]$ and by Theorem 7.1.1 equality (7.1.4) holds, and also

$$k_m^{(s)}\left(x + \frac{1}{2^{s-1}}\right) = (-1)^{\sigma([2^{s-1}(x+1)+1])} \cdot 2^{s(s+1)/2} k_{m-s}(-1 + 2\cdot\{2^{s-1}x\}) \qquad (7.1.12)$$

since $\{2^{s-1}(x + (1/2^{s-1}))\} = \{2^{s-1}x\}$. Then, (7.1.4) and (7.1.12) imply (7.1.11) and the corollary is proved.

**Remark**   Equality (7.1.11) is equivalent to

$$k_m^{(s)}(x) = (-1)^{\sigma([2^{s-1}(x+1)+1]) - \sigma([2^{s-1}(x+1)])} k_m^{(s)}\left(x + \frac{1}{2^{s-1}}\right). \qquad (7.1.13)$$

In the next theorem we need the notion of zero indicatrix.

A function $i_0(.): \mathbb{N}_0 \to \mathbb{N}_0$ is said to be a zero indicatrix in the binary system if the representation of the number $p \in \mathbb{N}_0$ in the binary system

$$p = \alpha_s \ldots \alpha_{q+1} \alpha_q \ldots \alpha_{1_{(2)}} = \sum_{v=1}^{s} \alpha_v \cdot 2^{v-1}$$

implies the equation $i_0(p) = q$; if, and only if, when $\alpha_{q+1} = 1$, $\alpha_q = \alpha_{q-1} = \ldots = \alpha_1 = 0$.

### Theorem 7.1.2

*Let $m = 2, 3, \ldots$; $s = 1, 2, \ldots, m-1$; $p \in \mathbb{Z}$, $I_{s,p} \overset{\text{def}}{=} [(p-1)/2^{s-1}, p/2^{s-1}]$ and $I_{s,p} \subset [-1, 1 - (1/2^{s-1})]$. Then*

$$k_m^{(s)}\left(x + \frac{1}{2^{s-1}}\right) = (-1)^{i_0(2^{s-1}+p)-1} k_m^{(s)}(x), \quad x \in I_{s,p}. \qquad (7.1.14)$$

### Proof

For $x = (p-1)/2^{s-1}$ or $x = p/2^{s-1}$ relation (7.1.14) follows from the elementary property 10* of the basic quasi-splines. This being so, let $x \in ((p-1)/2^{s-1}, p/2^{s-1})$. Then

$$[2^{s-1}(x+1)] = 2^{s-1} + p - 1, \quad [2^{s-1}(x+1)+1] = 2^{s-1} + p. \qquad (7.1.15)$$

The hypothesis $i_{s,p} \subset [-1, 1 - (1/2^{s-1})]$ implies

$$0 \leq 2^{s-1} + p - 1 < 2^{s-1} + p \leq 2^s - 1.$$

Therefore, if $i_0(2^{s-1} + p) = q$, then

$$2^{s-1} + p = \alpha_s \ldots \alpha_{q+2} 1 \underbrace{00 \ldots 0}_{q}{}_{(2)}$$

and

$$2^{s-1} + p - 1 = \alpha_s \ldots \alpha_{q+2} 0 \underbrace{11 \ldots 1}_{q}{}_{(2)}:$$

Hence

$$\sigma(2^{s-1} + p - 1) - \sigma(2^{s-1} + p) = q - 1 = i_0(2^{s-1} + p) - 1. \quad (7.1.16)$$

Then (7.1.11), (7.1.15) and (7.1.16) imply (7.1.14) and the theorem is proved.

Using Theorem 7.1.2, we can obtain a partition of the polynomials into linear combinations of shifts of basic quasi-splines.

### Theorem 7.1.3

*Let* $m = 2, 3, \ldots$; $s = 1, 2, \ldots, m-1$; $p \in \mathbb{Z}$, $I_{s,p} = [(p-1)/2^{s-1}, p/2^{s-1}]$ *and* $I_{s,p} \subset [-1, 1 - (1/2^{s-1})]$. *Then*

$$k_m(x) + (-1)^{i_0(2^{s-1}+p)} k_m\left(x + \frac{1}{2^{s-1}}\right) = \sum_{v=0}^{s-1} a_v \cdot \left(x - \frac{p-1}{2^{s-1}}\right)^v, \quad x \in I_{s,p} \quad (7.1.17)$$

*where* $i_0(.)$ *is the zero indicatrix in binary system and*

$$a_v = \frac{1}{v!}\left(k_m^{(v)}\left(\frac{p-1}{2^{s-1}}\right) + (-1)^{i_0(2^{s-1}+p)} k_m^{(v)}\left(\frac{p}{2^{s-1}}\right)\right), \quad v = 0, 1, \ldots, s-1. \quad (7.1.18)$$

### Proof

Equality (7.1.17) can be obtained from equality (7.1.14) by consecutive integration from $(p-1)/2^{s-1}$ to $x \in I_{s,p}$.

**Corollary 7.1.2**    *If* $m = 2, 3, \ldots$; $s = 1, 2, \ldots, m-1$, *then*

$$k_m(x) + (-1)^{s-1} k_m\left(\frac{1}{2^{s-1}} - x\right) = (-1)^{s-1} + \sum_{v=0}^{s-1} \frac{k_m^{(v)}(1/2^{s-1})}{v!}\left(x - \frac{1}{2^{s-1}}\right)^v$$

$$(7.1.19)$$

*for every* $x \in [0, 1/2^{s-1}]$.

Indeed, (7.1.19) is implied by (7.1.17) for $p = 0$ after the substitution of $x$ by $-x$ and taking into account the evenness of the basic quasi-splines, property 10* and $k_m(0) = 1$.

Equality (7.1.19) with $s = 1$ imply the partition of unity

$$k_m(x) + k_m(1 - x) = 1, \quad x \in [0, 1], \quad m = 2, 3, \ldots .$$

Again (7.1.19) for $s = 2$ implies the following partition of the first degree polynomial

$$k_m(x) - k_m(\tfrac{1}{2} - x) = \tfrac{1}{2} - 2x, \quad x \in [0, \tfrac{1}{2}], \quad m = 3, 4, \ldots . \tag{7.1.20}$$

For $s = 3$ we get the following partition of the second degree polynomial:
$$k_m(x) + k_m(\tfrac{1}{4} - x) = 1 + k_m(\tfrac{1}{4}) + x - 4x^2, \quad x \in [0, \tfrac{1}{4}], \quad m \geq 4.$$

and so on.

Let us note that according to Theorem 7.1.1, we have

$$k_m^{(v)}\left(\frac{1}{2^{s-1}}\right) = -2^{v(v+1)/2} \cdot k_{m-v}(-1 + 2^{v+1-s}) \tag{7.1.22}$$

for $m = 2, 3, \ldots; \; s = 1, 2, \ldots, m - 1; \; v = 1, 2, \ldots, s - 1, \; (v \geq 1!)$.

In particular,

$$k_m^{(s-1)}(1/2^{s-1}) = -2^{(s-1)s/2}, \quad s = 2, 3, \ldots, \quad m \geq 3 \tag{7.1.23}$$

$$k_m^{(s-2)}(1/2^{s-1}) = -2^{s(s-3)/2}, \quad s = 3, 4, \ldots, \quad m \geq 4. \tag{7.1.24}$$

(7.1.23) implies that the right-hand side of (7.1.19) is a polynomial of degree $s - 1$. This allows us to assert that every polynomial of degree $n (n \in \mathbb{N}_0)$ can be partitite in the segment $[0, 1/2^n]$ as a linear combination of the functions (with $m \geq n + 2$)

$$k_m(x), k_m(1 - x), k_m(\tfrac{1}{2} - x), \ldots, k_m((1/2^n) - x). \tag{7.1.25}$$

## 7.2   Monotonic sequences of basic quasi-splines

First, we will prove several lemmas.

### Lemma 7.2.1

*Let the functions $\varphi(x)$ and $f(x)$ belonging to the class A (see Section 6.1),* *satisfy the inequality*

$$\varphi(x) \leqq f(x), \quad 0 \leqq x \leqq \tfrac{1}{2}$$

*and the curves with equations $y = \varphi(x)$ $(-1 \leqq x \leqq 0)$ and $y = f(x)$, $-1 \leqq x \leqq 0$ be symmetric with respect to the point $(-\tfrac{1}{2}, \tfrac{1}{2})$. Then relations*

$$J(x) \overset{\text{def}}{=} \int_{2x-1}^{0} (f(x) - \varphi(x))\,dt \geq 0, \quad 0 \leqq x \leqq \tfrac{1}{2} \tag{7.2.1}$$

*and*

$$\sup\{J(x): 0 \leqq x \leqq \tfrac{1}{2}\} = J(\tfrac{1}{4}) \tag{7.2.2}$$

*hold true.*

### Proof

Let $x \in [0, \tfrac{1}{2}]$. Then $-1 \leq 2x - 1 \leq 0$. The integral $J(x)$ determines the area of the curvilinear trapezium, bounded by the curves $y = f(t)$, $y = \varphi(t)$, and the lines $t = 2x - 1$, $t = 0$ (figure 7.2.1).

From the hypothesis it follows that this trapezium for $x = 0$ consists of two equi-area parts, symmetric with respect to the point $(-\tfrac{1}{2}, \tfrac{1}{2})$. While $x$ varies from 0 to $\tfrac{1}{2}$, $t = 2x - 1$ varies from $-1$ to 0 and the area $J(x)$ of this

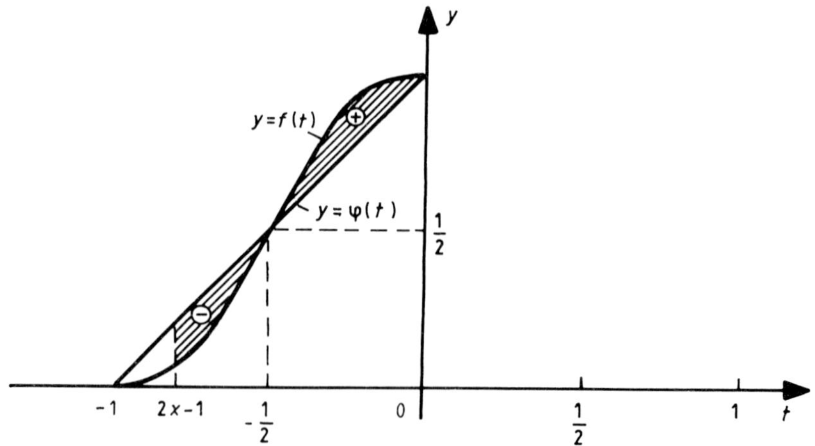


**Figure 7.2.1**   Graph of the function $J(x)$.

first trapezium varies from zero to its greatest value, which is attained from $t = -\frac{1}{2}$, i.e. for $2x - 1 = -\frac{1}{2} \Leftrightarrow x = \frac{1}{4}$, and then it decreases, and for $t = 2x - 1 = 0$ it attains again the value 0. Thus

$$J(x) \geq 0, \quad 0 \leq x \leq \tfrac{1}{2}$$

and

$$\sup\{ J(x) : 0 \leq x \leq \tfrac{1}{2}\} = J(\tfrac{1}{4}).$$

The lemma is proved.

## Lemma 7.2.2

*Let $m$ and $p$ be positive integers. Then*

$$\delta_p(x) \leq k_{m+p}(x) \leq \lambda_p(x), \quad 0 \leq x \leq \tfrac{1}{2} \tag{7.2.3}$$

*where $\delta_v(x)$, $k_v(x)$ and $\lambda_v(x)$ are the corresponding basic quasi-splines of order $v$ of the functions (6.1.7), $k(x) \in A$ and (6.1.3). The left-hand side of inequalities (7.1.3) is satisfied for $m = 0$ also.*

### Proof

For $p = 1$ relations (7.2.3) are proved in (6.3.2). Let (7.2.3) be true for some $p \in \mathbb{N}$. Then we will prove its validity for $p + 1$. More specifically, we will prove the inequality

$$k_{m+p+1}(x) \leq \lambda_{p+1}(x), \quad 0 \leq x \leq \tfrac{1}{2}. \tag{7.2.4}$$

The inequality

$$\delta_{p+1}(x) \leq k_{m+p+1}(x), \quad 0 \leq x \leq \tfrac{1}{2}$$

can be proved in the same way.

We use Lemma 7.2.1 with $\varphi(x) = k_{m+p}(x)$ and $f(x) = \lambda_p(x)$. Then, from the elementary property 8* of the basic quasi-splines it follows that:
(i) $\lambda_p(x) \in A$, $k_{m+p}(x) \in A$;
from the induction assumption we have
(ii) $k_{m+p}(x) \leq \lambda_p(x)$, $0 \leq x \leq \tfrac{1}{2}$; $m, p \in \mathbb{N}$;
from the elementary property 13* of the basic quasi-splines we obtain that:
(iii) the curves $y = k_{m+p}(x)$ $(-1 \leq x \leq 0)$ and $y = \lambda_p(x)$ $(-1 \leq x \leq 0)$ are symmetric with respect to the point $(-\tfrac{1}{2}, \tfrac{1}{2})$.

Hence, the hypothesis of Lemma 7.2.1 is satisfied. Then, from the assertion of this lemma it follows that the relation

$$0 \leqq \int_{2x-1}^{0} (\lambda_p(t) - k_{m+p}(t))\mathrm{d}t \leqq \int_{-1/2}^{0} (\lambda_p(t) - k_{m+p}(t))\mathrm{d}t, \quad 0 \leq x \leq \tfrac{1}{2} \quad (7.2.5)$$

is satisfied.

Then, from (6.1.2), elementary property 6*, (7.2.5) and from $0 \leq x \leq \tfrac{1}{2}$ it follows that

$$\lambda_{p+1}(x) - k_{m+p+1}(x) = \int_{2x-1}^{1} (\lambda_p(t) - k_{m+p}(t))\mathrm{d}t$$

$$= \int_{2x-1}^{0} (\lambda_p(t) - k_{m+p}(t))\mathrm{d}t \geqq 0, \quad 0 \leq x \leq \tfrac{1}{2}.$$

Thus the lemma is proved.

**Corollary 7.2.1** *If $m \in \mathbb{N}$, $p \in \mathbb{N}$, $0 \leqq x \leqq \tfrac{1}{2}$, then with the notation of the previous lemma*

$$0 \leqq k_{m+p}(x) - \delta_p(x) \leq \lambda_p(x) - \delta_p(x) \leqq \lambda_p(\tfrac{1}{4}) - \delta_p(\tfrac{1}{4}) \quad (7.2.6)$$

$$0 \leqq \lambda_p(x) - k_{m+p}(x) \leq \lambda_p(x) - \delta_p(x) \leqq \lambda_p(\tfrac{1}{4}) - \delta_p(\tfrac{1}{4}). \quad (7.2.7)$$

**Proof**

The first and the second of inequalities (7.2.6) follow from Lemma 7.2.1, applies to the functions $\varphi(x) = \delta_{p-1}(x)$ and $f(x) = \lambda_{p-1}(x)$.

Inequalities (7.2.7) can be proved in the same way.

The sequence

$$k_1(x), k_2(x), \ldots, k_m(x), \ldots, \quad 0 \leqq x \leqq \tfrac{1}{2} \quad (7.2.8)$$

of the basic quasi-splines is said to be [43] increasing or decreasing depending on whether

$$k_1(x) \leq k_2(x), \quad 0 \leq x \leq \tfrac{1}{2}, \tag{7.2.9}$$

or

$$k_1(x) \geq k_2(x), \quad 0 \leq x \leq \tfrac{1}{2}. \tag{7.2.10}$$

A sequence of basic quasi-splines (7.2.8) is said to be monotonic if it is increasing or decreasing.

Relations (6.1.6) and (6.1.10) imply that the sequences of basic quasi-splines of functions (6.1.3) and (6.1.7) are decreasing and increasing, respectively, i.e. there are monotonic sequences of basic quasi-splines.

### Lemma 7.2.3 ([43]).

*If the sequence (7.2.8) of basic quasi-splines is increasing or decreasing, then for every $m \in \mathbb{N}$ the inequality*

$$k_m(x) \leq k_{m+1}(x), \quad 0 \leq x \leq \tfrac{1}{2}, \tag{7.2.11}$$

*or*

$$k_m(x) \geq k_{m+1}(x), \quad 0 \leq x \leq \tfrac{1}{2} \tag{7.2.12}$$

*respectively, is satisfied.*

### Proof

We will prove inequality (7.2.11). Inequality (7.2.12) can be proved in the same way.

Let the sequence (7.2.8) of basic quasi-splines be increasing. By definition, this means, that (7.2.9) holds, i.e. inequality (7.2.11) with the $m = 1$ is true. Let it be true for some $m \in \mathbb{N}$. We are required to prove it for $m + 1$.

To this end, we apply Lemma 7.2.1 with $\varphi(x) = k_m(x)$ and $f(x) = k_{m+1}(x)$. From $m \in \mathbb{N}$ and the elementary property 8* it follows that:
(i) $k_m(x) \in A$, $k_{m+1}(x) \in A$;
from the induction assumption we have
(ii) $k_m(x) \leq k_{m+1}(x)$, $0 \leq x \leq \tfrac{1}{2}$;
finally, the elementary property 13* gives
(iii) the curves $y = k_m(x)$ $(-1 \leq x \leq 0)$ and $y = k_{m+1}(x)$ $(-1 \leq x \leq 0)$ are symmetric (each of them, separately) with respect to the point $(-\tfrac{1}{2}, \tfrac{1}{2})$. Using

Lemma 7.2.1, we obtain the inequality

$$\int_{2x-1}^{0} (k_{m+1}(t) - k_m(t))dt \geq 0, \quad 0 \leq x \leq \tfrac{1}{2}$$

which, along with (6.1.2), 6* and $0 \leq x \leq \tfrac{1}{2}$ gives

$$k_{m+2}(x) - k_{m+1}(x) = \int_{2x-1}^{1} (k_{m+1}(t) - k_m(t))dt = \int_{2x-1}^{0} (k_{m+1}(t) - k_m(t))dt \geq 0.$$

The lemma is proved.

## Lemma 7.2.4

*If the sequence (7.2.8) of basic quasi-splines is either increasing or decreasing, then for every $m - 1 \in \mathbb{N}$ and $p \in \mathbb{N}$*

$$\max\{k_{m+p}(x) - k_m(x) : 0 \leq x \leq \tfrac{1}{2}\} = k_{m+p}(\tfrac{1}{4}) - k_m(\tfrac{1}{4}) \qquad (7.2.13)$$

*or*

$$\min\{k_{m+p}(x) - k_m(x) : 0 \leq x \leq \tfrac{1}{2}\} = k_{m+p}(\tfrac{1}{4}) - k_m(\tfrac{1}{4}). \qquad (7.2.14)$$

*respectively.*

## Proof

We will prove the assertion of the lemma for an increasing sequence of basic quasi-splines. For a decreasing sequence the proof is similar. Thus, let sequence (7.2.8) be increasing. Then
(i) $k_{m-1}(x) \in A, \quad k_{m+p+1}(x) \in A, \quad m - 1 \in \mathbb{N}, \quad p \in \mathbb{N};$
Lemma 7.2.3 implies
(ii) $k_{m-1}(x) \leq k_{m-1+p}(x), \quad 0 \leq x \leq \tfrac{1}{2};$
finally, from property 13* we obtain
(iii) the curves $y = k_{m-1}(x) \ (-1 \leq x \leq 0)$ and $y = k_{m-1+p}(x) \ (-1 \leq x \leq 0)$ are symmetric (each of them, separately with respect to the point $(-\tfrac{1}{2}, \tfrac{1}{2})$.
   Then, by Lemma 7.2.1

$$\max_{0 \leq x \leq 1/2} \int_{2x-1}^{0} (k_{m-1+p}(t) - k_{m-1}(t))dt = \int_{-1/2}^{0} (k_{m-1+p}(t) - k_{m-1}(t))dt.$$

Since for $0 \leqq x \leqq \frac{1}{2}$

$$k_{m+p}(x) - k_m(x) = \int_{2x-1}^{0} (k_{m-1+p}(t) - k_{m-1}(t))dt$$

then the previous relation is equivalent to (7.2.13) and the lemma is proved.

**Corollary 7.2.2** *If the sequence (7.2.8) of basic quasi-splines is monotonic, then the sequence*

$$k_1(\tfrac{1}{4}), k_2(\tfrac{1}{4}), \ldots, k_m(\tfrac{1}{4}), \ldots$$

*is convergent.*

## Proof

Let, for example, the sequence (7.2.8) be increasing. By definition, this means that for each $x \in [0, \frac{1}{2}]$ $k_1(x) \leqq k_2(x)$. By Lemma 7.2.3, this implies that for every $m \in \mathbb{N}$ and for each $x \in [0, \frac{1}{2}]$

$$k_m(x) \leqq k_{m+1}(x)$$

and, in particular for $x = \frac{1}{4}$

$$k_m(\tfrac{1}{4}) \leqq k_{m+1}(\tfrac{1}{4}), \quad m \in \mathbb{N},$$

that is, the sequence $(k_m(\tfrac{1}{4}))_{m=1}^{\infty}$ is increasing. From the elementary properties 6*, 4*, 5* and 12* follow the inequalities $\frac{1}{2} \leqq k_m(\tfrac{1}{4}) \leqq 1$. Therefore, the sequence is bounded above and, hence, it is convergent.

**Corollary 7.2.3** *If sequence (7.2.8) of basic quasi-splines is monotonic, then it is convergent in the linear space $C[-1, 1]$ (and hence in $C(\mathbb{R})$).*

## Proof

The hypothesis of monotonicity of sequence (7.2.8) by Lemma 7.2.4 implies the relation

$$\|k_{m+p}(.) - k_m(.)\|_{C[0,1/2]} = |k_{m+p}(\tfrac{1}{4}) - k_m(\tfrac{1}{4})|; \quad m-1, \quad p \in \mathbb{N}.$$

Further, by 4*, 11*–13* there follows the relation

$$\|k_{m+p}(.)-k_m(.)\|_{C[-1,1]}=|k_{m+p}(\tfrac{1}{4})-k_m(\tfrac{1}{4})|, \quad m-1, \quad p\in\mathbb{N}. \quad (7.2.15)$$

Taking into account 2*, we note that in (7.2.15) instead of $C[-1,1]$ we can write $C(\mathbb{R})$.

Let $\varepsilon$ be an arbitrary positive number. By Corollary 7.2.2 one can find a number $m_0\in\mathbb{N}$ such that for every $m\geq m_0$ ($m\in\mathbb{N}$) and for each $p\in\mathbb{N}$ the inequality

$$|k_{m+p}(\tfrac{1}{4})-k_m(\tfrac{1}{4})|<\varepsilon \quad (7.2.16)$$

is satisfied.

From (7.2.15) and (7.2.16) it follows that the sequence $(k_m(x))_{m=1}^{\infty}$ is fundamental in the complete normed linear space $C[-1,1]$. Hence it is convergent in $C[-1,1]$. Let us denote its limit by $f(x)$. Then $f(x)\in C[-1,1]$ and

$$\lim_{m\to\infty}\|k_m(.)-f(.)\|_{C[-1,1]}=0. \quad (7.2.17)$$

Let us extend the definition of the function $f(x)$ on $x\in\mathbb{R}\backslash(-1,1)$ by $f(x)=0$. Then

$$\lim_{m\to\infty}\|k_m(.)-f(.)\|_{C(\mathbb{R})}=0. \quad (7.2.18)$$

The corollary is proved.


**Corollary 7.2.4**    *The sequence of the basic quasi-splines of the function (6.1.3) is uniformly convergent on* $\mathbb{R}$*:*

$$\lambda(x)\overset{\text{def}}{=}\lim_{m\to\infty}\lambda_m(x), \quad x\in\mathbb{R}. \quad (7.2.19)$$


This follows immediately from Corollary 7.2.3 and from the fact that the sequence of the basic quasi-splines of function (6.1.3) is decreasing.


**Corollary 7.2.5**    *The sequence of the basic quasi-splines of function (6.1.7) is uniformly convergent on* $\mathbb{R}$*:*

$$\delta(x)\overset{\text{def}}{=}\lim_{m\to\infty}\delta_m(x), \quad x\in\mathbb{R}. \quad (7.2.20)$$

**Corollary 7.2.6** *For the functions $\lambda(x)$ and $\delta(x)$, defined by (7.2.19) and (7.2.20) respectively,*

$$\delta(x) \leqq \lambda(x), \quad 0 \leqq x \leqq \tfrac{1}{2}. \tag{7.2.21}$$

Indeed, (7.2.3) implies the inequality

$$\delta_p(x) \leqq \lambda_p(x), \quad 0 \leqq x \leqq \tfrac{1}{2}, \quad p \in \mathbb{N}.$$

Letting $p \to \infty$, by (7.2.19) and (7.2.20) we obtain inequality (7.2.21).

Thus, both sequences of basic quasi-splines $(\lambda_m(x))_{m=1}^{\infty}$ and $(\delta_m(x))_{m=1}^{\infty}$, defined in Examples 6.1.1 and 6.1.2, respectively, are uniformly convergent on $\mathbb{R}$. The inequality (7.2.21) holds for their limit functions.

It is natural to ask the following question: can one assert that every sequence of basic quasi-splines is uniformly convergent on $\mathbb{R}$?

Let us assume the positive answer to this question and let us denote by $f(x)$ the limit function

$$f(x) = \lim_{m \to \infty} k_m(x), \quad x \in \mathbb{R}.$$

Since there are infinitely many functions, belonging to $A$ then (in the general case) we may expect that there are infinitely many limit functions for the corresponding sequence of basic quasi-splines, generated by these functions.

Passing to limit $m \to \infty$ in the elementary properties 2*, 4*–6* and in (6.1.2) suggests that the question just posed leads to the following problem.

---

### Problem C

Find all real functions $f : \mathbb{R} \to \mathbb{R}$ which satisfy the conditions:
(a) $f(x) = 0, \quad x \in \mathbb{R} \backslash (-1, 1)$;
(b) $f(-x) = f(x), \quad x \in \mathbb{R}$;
(c) $-1 \leqq x_1 < x_2 \leqq 0 \Rightarrow f(x_1) \leqq f(x_2)$;
(d) $f(0) = 1$;
(e) $f(x) = \int_{2x-1}^{2x+1} f(t)\,dt, \quad x \in \mathbb{R}$.

---

## 7.3 Properties of the solutions of Problem C

### Lemma 7.3.1

*If a function $f(x)$ is a solution of Problem C, then it belongs to the class $A$ and it coincides with each of its basic quasi-splines:*

$$f(x) \in A; \, f(x) = f_m(x), \quad x \in \mathbb{R}, \quad m \in \mathbb{N}_0. \tag{7.3.1}$$

## Proof

Let the function $f(x)$ be a solution of Problem C. From the conditions $(a)$–$(c)$ of this problem follow the conditions $(a)$–$(c)$ for belonging to the class $A$, and from conditions $(d)$ and $(e)$ of Problem C follow the relation

$$\int_{-1}^{1} f(x)\mathrm{d}x = f(0) = 1 \tag{7.3.2}$$

that is, it satisfies condition $(d)$ for the function $f(x)$ belongs to the class $A$. Hence $f(x) \in A$.

The second assertion of the lemma can be proved by induction on $m$. From (6.1.1) and (7.3.2) follow $f_0(x) = f(x)$, i.e. the assertion is true for $m = 0$. Let this assertion be true for $m = n \in \mathbb{N}_0$. Hence

$$f_n(x) = f(x), \quad x \in \mathbb{R}. \tag{7.3.3}$$

We will prove that it is true for $m = n + 1$ also, i.e.

$$f_{n+1}(x) = f(x), \quad x \in \mathbb{R}. \tag{7.3.4}$$

From (6.1.2), (7.3.3) and from condition $(e)$ of Problem C we obtain the consecutive equalities

$$f_{n+1}(x) = \int_{2x-1}^{2x+1} f_n(t)\mathrm{d}t = \int_{2x-1}^{2x+1} f(t)\mathrm{d}t = f(x), \quad x \in \mathbb{R}.$$

and from them (7.3.4) follows. Thus the lemma is proved.

Now, from Lemma 7.3.1 and from the properties of the basic quasi-splines the validity of the following assertion follows:

## Lemma 7.3.2

*If the function $f(x)$ is a solution of Problem C, then it has the following properties:*

$(i)$  $f(x) \geq 0, \quad x \in \mathbb{R};$
$(ii)$  $f(x) \in C^m(\mathbb{R}), \quad \forall m \in \mathbb{N}_0;$
$(iii)$  $f^{(s)}(x) = 2^{s(s+1)/2} \sum_{i_1=0}^{1} \cdots \sum_{i_s=0}^{1} (-1)^{i_1 + \ldots + i_s} \cdot f(2^s x + A_i(s))$

*where $x \in \mathbb{R}$, $s \in \mathbb{N}$, $i = (i_1, i_2, \ldots, i_s)$, and*

$$A_i(s) = \sum_{j=1}^{s} (-1)^{i_j} \cdot 2^{s-j}; \qquad (7.3.5)$$

*(iv) $f^{(s)}(p/2^v) = 0$, $p \in \mathbb{Z}$, $s \in \mathbb{N}$, $v = 0, 1, \ldots, s - 1$;*
*(v) $\|f^{(s)}(.)\|_C = 2^{s(s+1)/2} \|f(.)\|_C$, $s \in \mathbb{N}_0$;*
*(vi) $f(x) + f(1 - x) = 1$, $x \in [0, 1]$;*
*(vii) $f(-\tfrac{1}{2}) = f(\tfrac{1}{2}) = \tfrac{1}{2}$.*
*(viii) the function $f(x)$ is concave in the segment $[-\tfrac{1}{2}, \tfrac{1}{2}]$;*
*(ix) for every $m \in \mathbb{N}_0$ the inequalities*

$$\delta_m(x) \leq f(x) \leq \lambda_m(x), \quad 0 \leq x \leq \tfrac{1}{2} \qquad (7.3.6)$$

*hold, and also (again for $0 \leq x \leq \tfrac{1}{2}$) the inequalities:*

$$\begin{cases} 0 \leq \lambda_m(x) - f(x) \leq \lambda_m(x) - \delta_m(x) \leq \lambda_m(\tfrac{1}{4}) - \delta_m(\tfrac{1}{4}) \\ 0 \leq f(x) - \delta_m(x) \leq \lambda_m(x) - \delta_m(x) \leq \lambda_m(\tfrac{1}{4}) - \delta_m(\tfrac{1}{4}); \end{cases} \qquad (7.3.7)$$

*(x) if $m \in \mathbb{N}$ and $0 \leq x \leq \tfrac{1}{2}$, then*

$$0 \leq \lambda(x) - f(x) \leq \lambda_m(x) - \delta_m(x)$$

$$0 \leq f(x) - \delta(x) \leq \lambda_m(x) - \delta_m(x);$$

*(xi) if $\varphi(x)$ and $f(x)$ are two solutions of Problem C, satisfying the condition*

$$\varphi(x) \leq f(x), \quad 0 \leq x \leq \tfrac{1}{2} \qquad (7.3.8)$$

*then*

$$\int_{2x-1}^{1} (f(t) - \varphi(t)) dt = \int_{2x-1}^{0} (f(t) - \varphi(t)) dt \geq 0, \quad 0 \leq x \leq \tfrac{1}{2} \qquad (7.3.9)$$

$$\max\{f(x) - \varphi(x) : 0 \leq x \leq \tfrac{1}{2}\} = f(\tfrac{1}{4}) - \varphi(\tfrac{1}{4}) \qquad (7.3.10)$$

$$\|f(.) - \varphi(.)\|_{C[-1,1]} = f(\tfrac{1}{4}) - \varphi(\tfrac{1}{4}). \qquad (7.3.11)$$

### Proof

Taking into account that each solution of Problem C coincides with each
of its basic quasi-splines (Lemma 7.3.1), we see that properties (i)–(viii) of

these solutions follow immediately from the elementary properties of the basic quasi-splines with the corresponding numbers 3*, 7*, 9*, 10*, (7.1.4), 13*, 12*, Lemma 6.3.1. For $m = 0$ property (ix) can be checked directly, but for $m \in \mathbb{N}$ it follows from Lemma 7.2.2. Equalities (7.3.7) follow from Corollary 7.2.1.

From Lemma 7.2.2 and from the fact that the sequences of basic quasi-splines $(\lambda_m(x))_{m=1}^{\infty}$ and $(\delta_m(x))_{m=1}^{\infty}$ are decreasing and increasing, the corresponding inequalities (for $0 \leq x \leq \frac{1}{2}$) are:

$$\delta_p(x) \leq \delta_{m+p}(x) \leq f_{m+p+n}(x) \leq \lambda_{m+p}(x) \leq \lambda_p(x), \quad (m, n, p \in \mathbb{N}). \quad (7.3.12)$$

Using Lemma 7.3.1, by passing to limit for $m \to \infty$, we obtain the inequalities

$$\delta_p(x) \leq \delta(x) \leq f(x) \leq \lambda(x) \leq \lambda_p(x), \quad p \in \mathbb{N}, \quad 0 \leq x \leq \frac{1}{2}, \quad (7.3.13)$$

which are equivalent to (7.3.10).

Property (xi) follows from Lemma 7.2.1, Hypothesis (b) and Properties (vi) and (vii) of the solutions of Problem C.

### Theorem 7.3.1

*If $\varphi(x)$ and $f(x)$ are two solutions of Problem C, satisfying condition (7.3.8), then they coincide as follows:*

$$\varphi(x) = f(x), \quad x \in \mathbb{R}.$$

### Proof

It remains only to show that

$$\varphi(x) = f(x), \quad x \in [0, \tfrac{1}{2}], \quad (7.3.14)$$

since the remaining will follow from Hypothesis (a) of Problem C and from the symmetry of its solutions with respect to the axis $0y$ and with respect to the points $(-\frac{1}{2}, \frac{1}{2})$ and $(\frac{1}{2}, \frac{1}{2})$.

Thus, we are to prove (7.3.14). Let us assume the contrary: there exists at least one point $x_0 \in [0, \frac{1}{2}]$, such that $\varphi(x_0) \neq f(x_0)$. Then from (7.3.8) it follows that

$$f(x_0) - \varphi(x_0) > 0$$

which, combined with (7.3.10) gives

$$f(\tfrac{1}{4}) - \varphi(\tfrac{1}{4}) > 0. \tag{7.3.15}$$

Next, (7.3.15), (*e*), (7.3.9)–(7.3.11) give

$$0 < f(\tfrac{1}{4}) - \varphi(\tfrac{1}{4}) = \int\limits_{-1/2}^{0} (f(t) - \varphi(t))\mathrm{d}t$$

$$\leqq \|f(.) - \varphi(.)\|_{C[-1,1]} \cdot (0 - (-\tfrac{1}{2})) = \tfrac{1}{2}(f(\tfrac{1}{4}) - \varphi(\tfrac{1}{4}))$$

Thus giving the contradictory inequality

$$0 < f(\tfrac{1}{4}) - \varphi(\tfrac{1}{4}) \leqq \tfrac{1}{2}(f(\tfrac{1}{4}) - \varphi(\tfrac{1}{4})).$$

The theorem is proved.

## Theorem 7.3.2

*Problem C has a solution.*

## Proof

Let us pass to limit $m \to \infty$ in the elementary properties 2*, 4*–6* and in the definition equality (6.1.2) for the basic quasi-splines of function (6.1.3). Due to the uniform convergence of the function sequence $(\lambda_m(x))_{m=1}^{\infty}$ and the definition equality (7.2.19), we obtain the relations
(*a*) $\lambda(x) = 0, \quad x \in \mathbb{R} \setminus (-1, 1)$;
(*b*) $\lambda(-x) = \lambda(x), \quad x \in \mathbb{R}$;
(*c*) $-1 \leqq x_1 < x_2 \leqq 0 \Rightarrow \lambda(x_1) \leqq \lambda(x_2)$;
(*d*) $\lambda(0) = 1$;
(*e*) $\lambda(x) = \int_{2x-1}^{2x+1} \lambda(t)\mathrm{d}t, \quad x \in \mathbb{R}$.
They show that the function $\lambda(x)$, defined in (7.2.19), is a solution of Problem C. The theorem is proved.

---

**Remark.** In the same way we may show that the function $\delta(x)$, defined by (7.2.20), is also a solution of Problem C.

---

## Theorem 7.3.3

*The solution of Problem C is unique.*

### Proof

We know that the function $\lambda(x)$, defined in (7.2.19) is a solution of Problem C. Let $f(x)$ be an arbitrary solution of the same problem. According to Property $(x)$, Lemma 7.3.2 (see (7.3.13)) then

$$f(x) \leq \lambda(x), \quad 0 \leq x \leq \tfrac{1}{2}$$

which, by Theorem 7.3.1, is enough to assert that for each $x \in \mathbb{R}$ that $f(x) = \lambda(x)$.

The theorem is proved.

**Corollary 7.3.1**  *The function $\delta(x)$, defined in (7.2.20) for each $x \in \mathbb{R}$ coincides with the function $\lambda(x)$, defined in (7.2.19):*

$$\delta(x) = \lambda(x), \quad x \in \mathbb{R}.$$

From now on, we will denote the unique solution of Problem C by $\lambda(x)$.

Let $k(x)$ be an arbitrary function of the class $A$ and $k_m(x)$ and $m \in \mathbb{N}_0$ be its corresponding quasi-spline of order $m$. At the end of the previous section the question of the convergence of the function sequence $(k_m(x))_{m=1}^{\infty}$ was stated. The answer to this question is given in the next theorem.

## Theorem 7.3.4

*Every sequence of basic quasi-splines $(k_m(x))_{m=1}^{\infty}$ is uniformly convergent on $\mathbb{R}$ to $\lambda(x)$:*

$$\lim_{m \to \infty} k_m(x) = \lambda(x), \quad x \in \mathbb{R}. \tag{7.3.16}$$

### Proof

All will be settled, provided we show that

$$\lim_{m \to \infty} k_m(x) = \lambda(x), \quad x \in \mathbb{R} \tag{7.3.17}$$

uniformly over the segment $[0, \frac{1}{2}]$, since the remaining part follows from the relations:

(i) $k_m(x) = \lambda(x) = 0 \quad x \in \mathbb{R} \setminus (-1, 1), \quad m \in \mathbb{N}_0$ and

(ii) the symmetry of $\lambda(x)$ and $k_m(x)$, $m \in \mathbb{N}$ with respect to the axis $0y$ and with respect to the point $(\frac{1}{2}, \frac{1}{2})$.

Thus, we are to prove (7.3.17).

Let $\varepsilon > 0$ be an arbitrary positive number. One can find $m_0 \in \mathbb{N}$ such that for every $m \geq m_0$ $(m \in \mathbb{N})$ and for each $x \in [0, \frac{1}{2}]$ such that

$$|\lambda(x) - \lambda_m(x)| < \varepsilon/2 \tag{7.3.18}$$

since we have

$$\lim_{m \to \infty} \lambda_m(x) = \lambda(x)$$

uniformly over the segment $[0, \frac{1}{2}]$. Furthermore, we have

$$|\lambda(x) - \delta_m(x)| < \varepsilon/2 \tag{7.3.19}$$

since

$$\lim \delta_m(x) = \delta(x) = \lambda(x)$$

uniformly over $[0, \frac{1}{2}]$.

From the other side, Lemma 7.2.2 and (7.5.6) imply the inequality

$$|\lambda(x) - k_m(x)| \leq \lambda_{m-1}(x) - \delta_{m-1}(x), \quad 0 \leq x \leq \tfrac{1}{2}, \quad m = 2, 3, \dots . \tag{7.3.20}$$

From (7.3.18)–(7.3.20) for every $m \geq \max(2, m_0 + 1)$ and for each $x \in [0, \frac{1}{2}]$ it follows that

$$|\lambda(x) - k_m(x)| \leq \lambda_{m-1}(x) - \delta_{m-1}(x)$$

$$= \lambda_{m-1}(x) - \lambda(x) + \lambda(x) - \delta_{m-1}(x) < \frac{\varepsilon}{2} + \frac{\varepsilon}{2} = \varepsilon$$

that is, uniformly over $[0, \frac{1}{2}]$ we have

$$\lim_{m \to \infty} k_m(x) = \lambda(x).$$

The theorem is proved.

Therefore, the function $\lambda(x)$ is a universal limit of every sequence of basic quasi-splines.

From inequalities (7.3.6) follow functions $\lambda_m(x)$ and $\delta_m(x)$, defined in Examples 6.1.1 and 6.1.2, respectively, are one-sided monotonous approximations, from above and below, of the function $\lambda(x)$ over the segment $[0, \frac{1}{2}]$.

The problem for determining the error made in approximating the function $\lambda(x)$ by basic quasi-splines, is a natural one.

Relations (7.3.20), (7.3.6) and (7.3.7) and the symmetry of $\lambda(x)$ and of the basic quasi-splines with respect to the axis $0y$ and with respect to the point $(\frac{1}{2}, \frac{1}{2})$ imply the following assertion.

### Theorem 7.3.5

*For every function $k(x) \in A$*

$$\|\lambda(.) - k_{m+1}(.)\|_{C[-1,1]} \leqq \lambda_m(\tfrac{1}{4}) - \delta_m(\tfrac{1}{4}), \quad m \in \mathbb{N} \tag{7.3.21}$$

$$\|\lambda(.) - \lambda_m(.)\|_{C[-1,1]} \leqq \lambda_m(\tfrac{1}{4}) - \delta_m(\tfrac{1}{4}), \quad m \in \mathbb{N}_0 \tag{7.3.22}$$

*and*

$$\|\lambda(.) - \delta_m(.)\|_{C[-1,1]} \leqq \lambda_m(\tfrac{1}{4}) - \delta_m(\tfrac{1}{4}), \quad m \in \mathbb{N}_0. \tag{7.3.23}$$

A rough estimate for the approximations of the function $\lambda(x)$ by basic quasi-splines is given in the following theorem.

### Theorem 7.3.6

*For every function $k(x) \in A$*

$$\|\lambda(.) - k_m(.)\|_{C[-1,1]} \leqq 1/2^m, \quad m \in 2, 3, \ldots \tag{7.3.24}$$

$$\|\lambda(.) - \lambda_m(.)\|_{C[-1,1]} \leqq 1/2^{m+1}, \quad m \in \mathbb{N}_0 \tag{7.3.25}$$

*and*

$$\|\lambda(.) - \delta_m(.)\|_{C[-1,1]} \leqq 1/2^{m+1}, \quad m \in \mathbb{N}_0. \tag{7.3.76}$$

The proof will be completed if we can prove that

$$\lambda_m(\tfrac{1}{4}) - \delta_m(\tfrac{1}{4}) \leq 1/2^{m+1}, \quad m \in \mathbb{N}_0 \tag{7.3.27}$$

since the next part follows from Theorem 7.3.5.

For $m = 0$ relation (7.3.27) follows from (6.1.3) and (6.1.7). Let (7.3.27) be true from some $m \in \mathbb{N}_0$. We will prove it for $m + 1$. First, by Lemma 7.2.1, we get

$$\|\lambda_m(.) - \delta_m(.)\|_{C[-1,1]} \leqq \lambda_m(\tfrac{1}{4}) - \delta_m(\tfrac{1}{4}), \quad m \in \mathbb{N}_0. \tag{7.3.28}$$

Next from (6.1.2), 6* (7.3.28) and the induction assumption, we get

$$0 \leqq \lambda_{m+1}(\tfrac{1}{4}) - \delta_{m+1}(\tfrac{1}{4}) = \int_{-1/2}^{1} (\lambda_m(t) - \delta_m(t))\,dt$$

$$= \int_{-1/2}^{0} (\lambda_m(t) - \delta_m(t))\,dt \leq (\lambda(\tfrac{1}{4}) - \delta_m(\tfrac{1}{4})) \cdot \frac{1}{2} \leqq \frac{1}{2^{m+1}} \cdot \frac{1}{2}$$

$$= \frac{1}{2^{m+2}}.$$

The theorem is proved.

Theorem 7.3.6 shows that the basic quasi-splines approximate the function $\lambda(x)$ over the segment $[-1, 1]$ with the same order, as the polynomials approximate the analytic functions ([7], p 288). Thus, the natural question arises: is the function $\lambda(x)$ analytic in the segment $[-1, 1]$?

In order to answer this question, we will make a preliminary consideration of the question for strictly increasing behaviour and for strong convexity of the function $\lambda(x)$.

### Lemma 7.3.3

*The function $\lambda(x)$ is strictly increasing in each interval of the form $[-1/2^n, -1/2^{n+1}]$, $n \in \mathbb{N}$.*

The proof proceeds by induction on $n$.

Let $n = 1$ and $x \in [-1/2, -1/2^2]$. Then $2x - 1 \in [-2, -3/2]$ and $2x + 1 \in [0, 1/2]$. This and the conditions $(a)$ and $(e)$ of Problem C imply the identify

$$\lambda(x) = \int_{-1}^{2x+1} \lambda(t)\,dt \tag{7.3.29}$$

which gives the equation

$$\lambda'(x) = 2 \cdot \lambda(2x + 1). \tag{7.3.30}$$

Since the function $\lambda(x)$ decreases over the segment $[0, \frac{1}{2}]$ and $\lambda(\frac{1}{2}) = \frac{1}{2}$, then (7.3.30) implies the inequality

$$\lambda'(x) = 2 \cdot \lambda(2x + 1) \geq 2 \cdot \lambda(\tfrac{1}{2}) = 1 > 0, \quad -1/2 \leq x \leq -1/2^2$$

which is sufficient to assert that the lemma is true for $n = 1$.

Let the assertion of the lemma be true for some $n \in \mathbb{N}$, i.e. let $\lambda(x)$ be strictly increasing in $[-1/2^n, -1/2^{n+1}]$. By the symmetry of the curve $y = \lambda(x)$ $(-1 \leq x \leq 1)$ with respect to the axis $0y$ the function is strictly decreasing over the segment $[1/2^{n+1}, 1/2^n]$, and by the symmetry of the curve $y = \lambda(x)$ $(0 \leq x \leq 1)$ with respect to the point $(\frac{1}{2}, \frac{1}{2})$ it is strictly decreasing over the segment $[1 - 1/2^n, 1 - (1/2^{n+1})]$.

We will prove that the assertion of the lemma is true for $n + 1 \in N$ also. Let $x \in (-1/2^{n+1}, -1/2^{n+2})$. Then $2x + 1 \in (1 - (1/2^n), 1 - (1/2^{n+1}))$ and $2x - 1 \in (-1 - (1/2^n), -1 - (1/2^{n+1}))$. Therefore, formulas (7.3.29) and (7.3.30) remain valid. From (7.3.30), $x \in (-1/2^{n+1}, -1/2^{n+2})$ (more specifically, from the relation $2x + 1 \in (1 - (1/2^n), 1 - (1/2^{n+1}))$ implies by it) and from the induction assumption for the strictly decreasing behaviour of $\lambda(x)$ in the interval $(1 - (1/2^n), 1 - (1/2^{n+1}))$ it follows that

$$\lambda'(x) = 2 \cdot \lambda(2x + 1) > 2 \cdot \lambda(1 - (1/2^{n+1})) \geq 0, \quad x \in (-1/2^{n+1}, -1/2^{n+2}).$$

$$\tag{7.3.31}$$

Next, let $-1/2^{n+1} \leq x_1 < x_2 \leq -1/2^{n+2}$. Then from the theorem of finite increments and from (7.3.31) we have the estimate

$$\lambda(x_2) - \lambda(x_1) = \lambda'(\eta) \cdot (x_2 - x_1) > 0$$

since

$$-1/2^{n+1} \leq x_1 < \eta < x_2 \leq -1/2^{n+2}.$$

The lemma is proved.

### Theorem 7.3.7

*The function $\lambda(x)$ is strictly increasing in the segment $[-1, 0]$ and strictly decreasing in the segment $[0, 1]$.*

The proof will be completed, provided we ran prove that the function $\lambda(x)$

is strictly increasing in the segment $[-\frac{1}{2}, 0]$, since the remaining part will follow from Lemma 7.3.2 and Problem C.

Let $-\frac{1}{2} \leq x_1 < x_2 \leq 0$. By Archimedian principle, there exists a positive integer $n$ such that

$$-1/2^n \leq x_1 < -1/2^{n+1}. \tag{7.3.32}$$

Then, by Lemma 7.3.3

$$\lambda(x_1) < \lambda(-1/2^{n+1}). \tag{7.3.33}$$

For the number $x_2$ two cases are possible:

(i) $\quad x_2 \leq -1/2^{n+1}$ $\quad$ and $\quad$ (ii) $\quad -1/2^{n+1} < x_2$.

In the first case, the inequalities

$$-1/2^n \leq x_1 < x_2 \leq -1/2^{n+1}$$

are satisfied, and hence the assertion of the theorem follows from Lemma 7.3.3. In the second case from the condition $(c)$ of Problem C it follows that

$$\lambda(-1/2^{n+1}) \leq \lambda(x_2). \tag{7.3.34}$$

Relations (7.3.33) and (7.3.34) imply the inequality $\lambda(x_1) < \lambda(x_2)$, which shows that the theorem also remains valid in this case.
The proof is completed.

**Corollary 7.3.2** *For every* $x \in (-1, 1)$

$$\lambda(x) > 0 \tag{7.3.35}$$

*holds.*

**Corollary 7.3.4** *The function* $\lambda(x)$ *is strictly convex in the segments* $[-1, -\frac{1}{2}]$ *and* $[\frac{1}{2}, 1]$ *and strictly concave on the segment* $[-\frac{1}{2}, \frac{1}{2}]$.

**Proof**

It remains only to prove that $\lambda(x)$ is strictly concave over the segment $[0, \frac{1}{2}]$ since the remaining part follows from the symmetry of the curve $y = \lambda(x)$

$(0 \leq x \leq 1)$ with respect to the point $(\frac{1}{2}, \frac{1}{2})$ and on the curve $y = \lambda(x)$ $(-1 \leq x \leq 1)$ with respect to the ordinate axis.

Let $x \in [0, \frac{1}{2}]$, $y \in [0, \frac{1}{2}]$ and let e.g. $x < y$. Then

$$0 \leq x < (x + y)/2 < y \leq \tfrac{1}{2}$$

$$-1 \leq 2x - 1 < x + y - 1 < 2y - 1 \leq 0$$

$$1 \leq 2x + 1 < x + y + 1 < 2y + 1$$

and

$$\lambda((x + y)/2) - (\lambda(x) + \lambda(y))/2 = \tfrac{1}{2} \int_{x+y-1}^{2y-1} \lambda(t)\,dt - \tfrac{1}{2} \int_{2x-1}^{x+y-1} \lambda(t)\,dt$$

$$> \tfrac{1}{2}\lambda(x + y - 1) \cdot (y - x - (y - x)) = 0$$

since by Theorem 7.3.7, the function $\lambda(x)$ is strictly increasing in the segment $[-1, 0]$. The corollary is proved.

Further, using Lemma 7.3.1 and the results of Section 7.1, we will obtain several properties of the function $\lambda(x)$.

### Lemma 7.3.4

*If $s \in \mathbb{N}$ and $x \in \mathbb{R}$, then*

$$\lambda^{(s)}(x) = 2^{s(s+1)/2} \sum_{i=0}^{2^s - 1} (-1)^{\sigma(i)} \cdot \lambda(2^s(x + 1) - 1 - 2i) \qquad (7.3.36)$$

*where $\sigma(i)$ is the sum of the binary digits of the number $i$ (see (7.1.2)).*

The assertion follows from Lemmas 7.1.1 and 7.3.1.

### Theorem 7.3.8

*If $s \in \mathbb{N}$ and $x \in [-1, 1]$, then*

$$\lambda^{(s)}(x) = (-1)^{\sigma([2^{s-1}(x+1)])} \cdot 2^{s(s+1)/2} \cdot \lambda(-1 + 2 \cdot \{2^{s-1}x\}) \qquad (7.3.37)$$

where $\sigma(i)$ is the sum of the binary digits of the number $i$ and $[\alpha]$ and $\{\alpha\}$ denote the integer and fractional parts of the number $\alpha$, respectively.

The assertion follows directly from Theorem 7.1.1 and Lemma 7.3.1.

**Corollary 7.3.5** *If* $s\in\mathbb{N}$ *and* $x\in[-1, 1-(1/2^{s-1})]$, *then*

$$\lambda^{(s)}(x + (1/2^{s-1})) = (-1)^{\sigma([2^{s-1}(x+1)]-\sigma([2^{s-1}(x+1)+1]))}\cdot\lambda^{(s)}(x) \quad (7.3.38)$$

where $\sigma(i)$ is the sum of the binary digits of the number $i$ and $[\alpha]$ denotes the integer part of the number $\alpha$.

The assertion follows from Corollary 7.1.1 and Lemma 7.3.1.

### Theorem 7.3.9

*Let* $s\in\mathbb{N}, p\in\mathbb{Z}, I_{s,p} = [(p-1)/2^{s-1}, p/2^{s-1}]$ *and* $I_{s,p} \subset [-1, 1-(1/2^{s-1})]$. *Then for every* $x\in I_{s,p}$

$$\lambda^{(s)}(x + (1/2^{s-1})) = (-1)^{i_0(2^{s-1}+p)-1}\lambda^{(s)}(x) \qquad (7.3.39)$$

where $i_0(.)$ is the zero indicatrix in the binary system, determined in Section 7.1.

The assertion follows from Theorem 7.1.2 and Lemma 7.3.1.

### Theorem 7.3.10

*If* $s\in\mathbb{N}, p\in\mathbb{Z}$ *and* $I_{s,p} \subset [-1, 1-(1/2^{s-1})]$, *then for every* $x\in I_{s,p}$

$$\lambda(x) + (-1)^{i_0(2^{s-1}+p)}\cdot\lambda(x + (1/2^{s-1})) = \sum_{v=0}^{s-1} a_v\cdot\left(x - \frac{p-1}{2^{s-1}}\right)^v \quad (7.3.40)$$

where $i_0(.)$ is the zero indicatrix in the binary system, $I_{s,p}$ is the interval, defined in the previous theorem and

$$a_v = \frac{1}{v!}\left(\lambda^{(v)}\left(\frac{p-1}{2^{s-1}}\right) + (-1)^{i_0(2^{s-1}+p)}\lambda^{(v)}\left(\frac{p}{2^{s-1}}\right)\right), \quad v = 0, 1, \ldots, s-1. \quad (7.3.41)$$

The assertion follows from Theorem 7.1.3 and Lemma 7.3.1.

**Corollary 7.3.6**   *If $s\in\mathbb{N}$, then for every $x\in[0,1/2^{s-1}]$*

$$\lambda(x)+(-1)^{s-1}\cdot\lambda((1/2^{s-1})-x)=(-1)^{s-1}+\sum_{v=0}^{s-1}\frac{\lambda^{(v)}(1/2^{s-1})}{v!}\cdot(x-(1/2^{s-1}))^{v}.$$

$$(7.3.42)$$

In particular,

$$\lambda(x)-\lambda(\tfrac{1}{2}-x)=\tfrac{1}{2}-2x,\quad x\in[0,\tfrac{1}{2}] \tag{7.3.43}$$

$$\lambda(x)+\lambda(\tfrac{1}{4}-x)=1+\lambda(\tfrac{1}{4})+x-4x^{2},\quad x\in[0,\tfrac{1}{4}] \tag{7.3.44}$$

and so on.

Let us note that according to (7.3.37)

$$\lambda^{(v)}(1/2^{s-1})=-2^{v(v+1)/2}\cdot\lambda(-1+2^{v+1-s}),\quad s\in\mathbb{N},\quad v=1,2,\ldots,s-1. \tag{7.3.45}$$

In particular,

$$\lambda^{(s-1)}(1/2^{s-1})=-2^{(s-1)s/2},\quad s=2,3,\ldots \tag{7.3.46}$$

$$\lambda^{(s-2)}(1/2^{s-1})=-2^{s(s-3)/2},\quad s=3,4,\ldots. \tag{7.3.47}$$

From (7.3.46) it follows that the right-hand-side of (7.3.42) is a polynomial of degree $s-1$.

We will now prove a proposition, which is important for the solution of the question for the analysis of function $\lambda(x)$ in the segment $[-1,1]$.

### Lemma 7.3.5

*If $s\in\mathbb{N}$, $p\in\mathbb{Z}$ and $((p-1)/2^{s-1},p/2^{s-1})\subset[-1,1]$, then for every $x\in((p-1)/2^{s-1}, p/2^{s-1})$*

$$\lambda^{(s)}(x)\neq0. \tag{7.3.48}$$

### Proof

Let $x\in((p-1)/2^{s-1},p/2^{s-1})\subset[-1,1]$. Then (since $s\in\mathbb{N}$) Theorem 7.3.8 is applicable as follows:

$$\lambda^{(s)}(x)=(-1)^{\sigma([2^{s-1}(x+1)])}\cdot2^{s(s+1)/2}\cdot\lambda(-1+2\{2^{s-1}x\}). \tag{7.3.49}$$

From the other side, the condition

$$x\in\left(\frac{p-1}{2^{s-1}},\frac{p}{2^{s-1}}\right)\subset[-1,1]$$

implies

$$-1<-1+2\cdot\{2^{s-1}x\}<1$$

which by Corollary 7.3.2 allows us to assert that

$$\lambda(-1+2\cdot\{2^{s-1}x\})>0 \tag{7.3.50}$$

Then (7.3.49) and (7.3.50) imply (7.3.48) and the lemma is proved.

**Corollary 7.3.7** *If $s\in\mathbb{N}, p\in\mathbb{Z}$ and $J_{s,p}\overset{\text{def}}{=}((p-1)/2^{s-1},p/2^{s-1})\subset[-1,1],$ then*

$$\text{sign}\{\lambda^{(s)}(x):x\in J_{s,p}\}=(-1)^{\sigma(2^{s-1}+p-1)} \tag{7.3.51}$$

*where $\sigma(i)$ is the sum of the binary digits of the number $i$, and*

$$\text{sign } x\begin{cases} 1 & \text{for } x>0 \\ 0 & \text{for } x=0. \\ -1 & \text{for } x<0 \end{cases}$$

**Proof**

Equation (7.3.51) follows from (7.3.49), (7.3.50) and from the fact that under the hypothesis of the corollary. $[2^{s-1}(x+1)]=2^{s-1}+p-1.$

## 7.4  Non-analyticity of the function $\lambda(x)$

A function $f(x)$ is said to be analytic (see, e.g. [7], p 216) in an interval $(a,b)$ when there exists a positive number $r$, such that for every $x_0\in(a,b)$ to exist a power series

$$\sum_{k=0}^{\infty} C_k(x_0)\cdot(x-x_0)^k \tag{7.4.1}$$

is converging for every $x \in (x_0 - r, x_0 + r)$ and each $x \in (a, b) \cap (x_0 - r, x_0 + r)$ has the representation

$$f(x) = \sum_{k=0}^{\infty} C_k(x_0) \cdot (x - x_0)^k. \tag{7.4.2}$$

If $r = +\infty$ the function $f(x)$ is said to be entire.

### Theorem 7.4.1

*The function $\lambda(x)$ is non-analytic in each interval $(a, b), a < b, (a, b) \subset (-1, 1)$.*

### Proof

Let us assume the contrary: that there exists at least an interval $(a, b), a < b$, $(a, b) \subset (-1, 1)$ in which the function $\lambda(x)$ is analytic. This means the following: there exists a positive number $r$, such that for every $x_0 \in (a, b)$ one can find a power series of the form (7.4.1), which is convergent for every $x \in (x_0 - r, x_0 + r)$ and such that for every $x \in (a, b) \cap (x_0 - r, x_0 + r)$ the equality

$$\lambda(x) = \sum_{k=0}^{\infty} C_k(x_0) \cdot (x - x_0)^k \tag{7.4.3}$$

holds.

Let us choose the number $s \in \mathbb{N}$ large enough such that

$$1/2^{s-1} < b - a. \tag{7.4.4}$$

According to the Archimedian principle, this choice is possible since $a < b$. Next, we denote

$$p - 1 = [a \cdot 2^{s-1}] \tag{7.4.5}$$

where $[\alpha]$ is the integer part of the number $\alpha$. Therefore,

$$a \cdot 2^{s-1} < p \Leftrightarrow a < p/2^{s-1}. \tag{7.4.6}$$

Then (7.4.4) and (7.4.5) imply the relations

$$\frac{p}{2^{s-1}} = \frac{p-1}{2^{s-1}} + \frac{1}{2^{s-1}} \leq a + \frac{1}{2^{s-1}} < a + b - a = b$$

which together with (7.4.6) give $p/2^{s-1} \in (a, b)$.

Further, from (7.4.3) we obtain that for every $x \in (\alpha, \beta)$

$$\lambda(x) = \sum_{k=0}^{\infty} C_k\left(\frac{p}{2^{s-1}}\right) \cdot \left(x - \frac{p}{2^{s-1}}\right)^k \tag{7.4.7}$$

where $\quad \alpha = \max(a, (p/2^{s-1}) - r) \geqq a > -1, \quad \beta = \min(b(p/2^{s-1}) + r) \leqq b < 1,$ $\alpha < \beta$.

By differentiation of (7.4.7), we can determine the coefficients of the series (7.4.7) as follows:

$$k! C_k(p/2^{s-1}) = \lambda^{(k)}(p/2^{s-1}), \quad k \in \mathbb{N}_0. \tag{7.4.8}$$

From Property (iv) of Lemma 7.3.2 we know that $\lambda^{(k)}(p/2^{s-1}) = 0$ for every positive integer $k > s - 1$ provided $p \in \mathbb{Z}$. That is why (7.4.7) and (7.4.8) imply the equality

$$\lambda(x) = \sum_{k=0}^{s-1} C_k(p/2^{s-1}) \cdot (x - (p/2^{s-1}))^k \tag{7.4.9}$$

for every $x \in (\alpha, \beta)$. Hence

$$\lambda^{(s)}(x) = 0 \tag{7.4.10}$$

for every $x \in (\alpha, \beta)$.

From $p/2^{s-1} \in (a, b)$ and $p/2^{s-1} \in ((p/2^{s-1}) - r, (p/2^{s-1}) + r)$ it follows $p/2^{s-1} \in (\alpha, \beta)$. Hence

$$\Delta \overset{\text{def}}{=} (\alpha, \beta) \cap \left(\frac{p-1}{2^{s-1}}, \frac{p}{2^{s-1}}\right) \neq \varnothing.$$

Let $x_1 \in \Delta$. Then, by Lemma 7.3.5

$$\lambda^{(s)}(x) \neq 0 \tag{7.4.11}$$

for every $x \in ((p-1)/2^{s-1}, p/2^{s-1})$.

Then (7.4.10) and (7.4.11) imply the contradictory relations

$$\lambda^{(s)}(x_1) = 0 \quad \text{and} \quad \lambda^{(s)}(x_1) \neq 0.$$

The theorem is proved.

Therefore, the answer to the question, posed in the previous section, is negative.

# 7.5   Localization of the zeros of the derivatives of the function $\lambda(x)$

In Property (iv) of Lemma 7.3.2 it was proved that

$$\lambda^{(s)}(p/2^v) = 0, \quad p \in \mathbb{Z}, \quad s \in \mathbb{N}, \quad v = 0, 1, 2, \ldots, s-1.$$

This naturally begs the question about the localization of the zeros of $\lambda^{(s)}(x)$. Lemma 7.3.5 allows us to answer this question.

## Theorem 7.5.1

*If $s \in \mathbb{N}$, then all the zeros of the function $\lambda^{(s)}(x)$ belonging to the segment $[-1, 1]$ are given by the formula*

$$x = \frac{p}{2^{s-1}} \tag{7.5.1}$$

*where the number $p \in \mathbb{Z}$ satisfies*

$$|p| \leq 2^{s-1}. \tag{7.5.2}$$

## Proof

Property (iv) of Lemma (7.3.2 implies that the numbers (7.5.1), (7.5.2) are zeros of $\lambda^{(s)}(x)$ belonging to the segment $[-1, 1]$. It remains to prove that these are all the zeros of $\lambda^{(s)}(x)$ in the segment $[-1, 1]$.

Let $x_0$ be an arbitrary number from the segment $[-1, 1]$ different from $p/2^{s-1}$, where $p \in \mathbb{Z}$ and $|p| \leq 2^{s-1}$. Then $x_0 \in (-1, 1)$ since if $x_0 = +1$ or $x_0 = -1$, then we would have $x_0 = \pm 2^{s-1}/2^{s-1}$ which contradicts the hypothesis. Thus

$$-1 < x_0 < 1, \quad x_0 \neq p/2^{s-1}.$$

We denote $q - 1 = [x_0 \cdot 2^{s-1}]$, where $[\alpha]$ is the integer part of the number $\alpha$. Hence $q - 1 \leq x_0 \cdot 2^{s-1} < q$.

Moreover, this and $x_0 \neq p/2^{s-1}$ imply

$$-2^{s-1} \leq q - 1 < x_0 2^{s-1} < q \leq 2^{s-1}$$

or, which is the same,

$$x_0 \in \left( \frac{q-1}{2^{s-1}}, \frac{q}{2^{s-1}} \right) \subset [-1, 1], \quad q \in \mathbb{Z}, \quad s \in \mathbb{N}.$$

By Lemma 7.3.5, we have $\lambda^{(s)}(x_0) \neq 0$. The theorem is proved.

Thus it was shown that all the zeros of the function $\lambda^{(s)}(x)$, $s \in \mathbb{N}$ which are in the segment $[-1, 1]$ are given by (7.5.1), (7.5.2) and (counted once) are exactly $2^s + 1$ in number. Each such zero has infinite multiplicity.

## 7.6  Tabulating, integral representation and central moments of the function $\lambda(x)$

Theorem 7.3.5 and 7.3.6 allow us to tabulate the function $\lambda(x)$. This option is realized in [54].

From the conditions of Problem C and from Lemma 7.3.2 we know that $\lambda(0) = 1$, $\lambda(x) = 0$ for $x \in \mathbb{R} \setminus (-1, 1)$, $\lambda(x)$ is an even function, and the graph of the curve $y = \lambda(x)$ $(0 \le x \le 1)$ is symmetric with respect to the point $(\frac{1}{2}, \frac{1}{2})$ and $\lambda(\frac{1}{2}) = \frac{1}{2}$.

Hence, it is sufficient to find its values under the restriction

$$0 < x < 0.5. \tag{7.6.1}$$

We will use the basic quasi-splines of the functions (6.1.3) and (6.1.7) and the inequalities (7.3.6), (7.3.7) and (7.3.21)–(7.3.26) as approximation aggregates and as a tool for estimation, respectively.

From (7.6.1) the inequalities

$$0 < 1 - 2x < 1, \quad 1 < 2x + 1 \tag{7.6.2}$$

follow. They allow us to write formula (6.1.2) in the form

$$k_m(x) = 0.5 + \int_0^{1-2x} k_{m-1}(t)\,dt, \quad x \in (0, \tfrac{1}{2}). \tag{7.6.3}$$

We need the values

$$k_m(x_i) \quad \text{for } x_i = (i-1)\cdot h \quad \text{when } h = 10^{-3}, \quad i = \overline{2,500}. \tag{7.6.4}$$

In order to calculate them we will use the iteration procedure (7.6.3). We will calculate the integral in (7.6.3) by the trapezium quadrature formula

$$\int_0^{1-2x_i} k_{m-1}(t)dt = \frac{h}{2}\left( k_{m-1}(x_1) + k_{m-1}(1-2x_i) + 2\cdot \sum_{j=2}^{1002-2i} k_{m-1}(x_j) \right) + R(k_{m-1})$$

(7.6.5)

with the same step and with an estimate of the remainder term

$$|R(k_{m-1})| \leqq \frac{1-2x_i}{12}\cdot h^2\cdot M_2 \quad M_2 = \|k_{m-1}\|_{C[0,1/2]}.$$  (7.6.6)

For $m > 3$ from (7.6.1) and (7.6.6) it follows that

$$|R(k_{m-1})| \leqq \frac{2}{3}\cdot 10^{-6}$$  (7.6.7)

since $M_2 \leqq 8$ (see (7.1.4)).

The method is to be applied to the functions $\lambda_m(x)$ and $\delta_m(x)$. The quadrature formulas for $\lambda_m(x_i)$ and $\delta_m(x_i)$ are denoted by $\tilde{\lambda}_m(x_i)$ and $\tilde{\delta}_m(x_i)$. respectively. For an approximate value of $\lambda(x_i)$ we accept $\tilde{\lambda}_m(x_i)$, and the number $m$ of the iterations is to be determined by the condition

$$|\tilde{\lambda}_m(\tfrac{1}{4}) - \tilde{\delta}_m(\tfrac{1}{4})| < 10^{-7}.$$  (7.6.8)

Then, by Theorem 7.3.5 and inequalities (7.6.7) and (7.6.8) we obtain the estimate

$$|\lambda(x_i) - \tilde{\lambda}_m(x_i)| \leqq |\lambda(x_i) - \lambda_m(x_i)| + |\lambda_m(x_i) - \tilde{\lambda}_m(x_i)$$
$$< \lambda_m(\tfrac{1}{4}) - \delta_m(\tfrac{1}{4}) + \tfrac{2}{3}10^{-6} \leqq |\lambda(\tfrac{1}{4}) - \tilde{\lambda}(\tfrac{1}{4})| + |\tilde{\lambda}(\tfrac{1}{4})$$
$$- \tilde{\delta}_m(\tfrac{1}{4})| + |\delta(\tfrac{1}{4}) - \tilde{\delta}(\tfrac{1}{4})| + \tfrac{2}{3}10^{-6} < 2 \times 10^{-6} + 10^{-7} = 2.1 \times 10^{-6}.$$

The condition (7.6.8) is satisfied for $m \geqq 11$ and it can be verified computationally. Hence $\tilde{\lambda}_{11}(x_i)$ gives the value of $\lambda(x_i)$ to at least five correct digits after the decimal point.

The computer calculations were carried out at the Computer Centre of Plovdiv University. Here we give the values of the function $\lambda(x)$ in the segment $[0,\tfrac{1}{2}]$ with the step 0.01 (table 7.6.1).

The graph of the function $\lambda(x)$ is shown in figure 7.6.1.

In order to obtain the integral representation of the function $\lambda(x)$ we use the Fourier cosine transform.

**Table 7.6.1**

| $x$ | $\lambda(x)$ | $x$ | $\lambda(x)$ | $x$ | $\lambda(x)$ |
|---|---|---|---|---|---|
| 0.00 | 1.00000 | 0.17 | 0.98526 | 0.34 | 0.80876 |
| 0.01 | 1.00000 | 0.18 | 0.98109 | 0.35 | 0.79164 |
| 0.02 | 1.00000 | 0.19 | 0.97621 | 0.36 | 0.77396 |
| 0.03 | 1.00000 | 0.20 | 0.97057 | 0.37 | 0.75578 |
| 0.04 | 1.00000 | 0.21 | 0.96416 | 0.38 | 0.73717 |
| 0.05 | 0.99998 | 0.22 | 0.95696 | 0.39 | 0.71820 |
| 0.06 | 0.99995 | 0.23 | 0.94896 | 0.40 | 0.69892 |
| 0.07 | 0.99986 | 0.24 | 0.94016 | 0.41 | 0.67939 |
| 0.08 | 0.99969 | 0.25 | 0.93056 | 0.42 | 0.65969 |
| 0.09 | 0.99939 | 0.26 | 0.92016 | 0.43 | 0.63986 |
| 0.10 | 0.99892 | 0.27 | 0.90896 | 0.44 | 0.61995 |
| 0.11 | 0.99820 | 0.28 | 0.89696 | 0.45 | 0.59998 |
| 0.12 | 0.99717 | 0.29 | 0.88416 | 0.46 | 0.58000 |
| 0.13 | 0.99578 | 0.30 | 0.87057 | 0.47 | 0.56000 |
| 0.14 | 0.99396 | 0.31 | 0.85621 | 0.48 | 0.54000 |
| 0.15 | 0.99164 | 0.32 | 0.84109 | 0.49 | 0.52000 |
| 0.16 | 0.98876 | 0.33 | 0.82526 | 0.50 | 0.50000 |

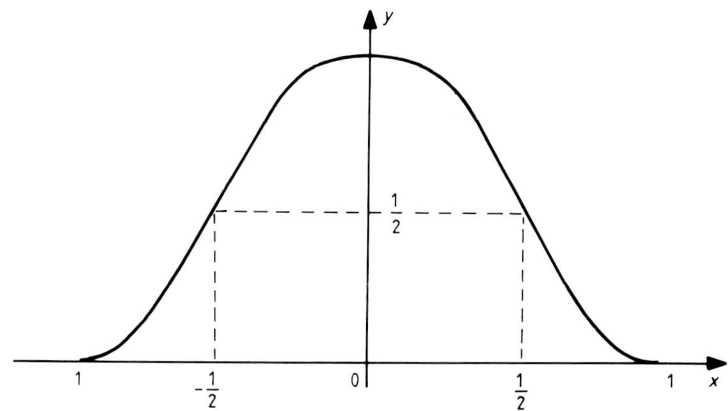

**Figure 7.6.1**   Graph of the function $\lambda(x)$ (the atomar function).

Let $\Lambda(t)$ be the image if $\lambda(x)$:

$$\Lambda(t) = \int_{-\infty}^{+\infty} \cos(tx) \cdot \lambda(x) \cdot dt = \int_{-1}^{1} \cos(tx) \cdot \lambda(x) \cdot dx. \qquad (7.6.9)$$

Then

$$\lambda(x) = \frac{1}{2\pi} \int\limits_{-\infty}^{+\infty} \cos(tx)\cdot\Lambda(t)\mathrm{d}t. \tag{7.6.10}$$

## Lemma 7.6.1

*For every $n\in\mathbb{N}$ and for each $t\in\mathbb{R}$*

$$|\Lambda(t/2^n) - 1| \leq \frac{t^2}{9\cdot 2^{2n+1}}. \tag{7.6.11}$$

## Proof

From (7.6.9) it follows that

$$\Lambda(t/2^n) = \int\limits_{-1}^{1} \cos(tx/2^n)\cdot\lambda(x)\mathrm{d}x$$

whence for every fixed $t\in\mathbb{R}$ we get the estimate

$$|\Lambda(t/2^n) - 1| = \left| \int\limits_{-1}^{1} \cos(tx/2^n)\cdot\lambda(x)\mathrm{d}x - \int\limits_{-1}^{1} \lambda(x)\mathrm{d}x \right|$$

$$\leq \int\limits_{-1}^{1} |\cos(tx/2^n) - 1|\cdot\lambda(x)\mathrm{d}x \leq t^2/2^{2n+1}\cdot \int\limits_{-1}^{1} x^2\cdot\lambda(x)\mathrm{d}x.$$

Therefore, in order to prove the lemma, it is sufficient to prove first that

$$\mu_2 \overset{\text{def}}{=} \int\limits_{-1}^{1} x^2\cdot\lambda(x)\mathrm{d}x = 1/9. \tag{7.6.12}$$

From Condition (*e*) of Problem C, by interchanging the order of the integration we obtain

$$\mu_2 = \int_{-1}^{1} x^2 \cdot \lambda(x)dx = \int_{-1}^{1} x^2 \cdot \int_{2x-1}^{2x+1} \lambda(t)dt\,dx$$

$$= \int_{-1}^{1} \int_{(t-1)/2}^{(t+1)/2} x^2 \cdot \lambda(t)dx\,dt = \int_{-1}^{1} \lambda(t) \cdot \frac{6t^2+1}{24}\,dt = \frac{1}{4}\mu_2 + \frac{1}{12}$$

that is

$$\mu_2 = \frac{1}{4}\mu_2 + \frac{1}{12} \Leftrightarrow \mu_2 = \frac{1}{9}.$$

The lemma is proved.

## Lemma 7.6.2

*For every $n \in \mathbb{N}$ and for each fixed $t \in \mathbb{R}$ we have*

$$\left| \Lambda(t) - \prod_{k=1}^{n} \frac{\sin(t2^{-k})}{t \cdot 2^{-k}} \right| \leqq \frac{t^2}{9 \cdot 2^{2n+1}}. \tag{7.6.13}$$

## Proof

From (7.6.9) and from Condition (*e*) of Problem C by interchanging the order of integration for $t \neq 0$, we get

$$\Lambda(t) = \int_{-1}^{1} \cos(tx) \cdot \lambda(x)dx = \int_{-1}^{1} \cos(tx) \cdot \int_{2x-1}^{2x+1} \lambda(u)du\,dx$$

$$= \int_{-1}^{1} \int_{(u-1)/2}^{(u+1)/2} \cos(tx) \cdot \lambda(u)dx\,du = \frac{\sin(t/2)}{(t/2)} \cdot \int_{-1}^{1} \cos((t/2)u)\lambda(u)du$$

that is

$$\Lambda(t) = \frac{\sin(t/2)}{(t/2)} \cdot \Lambda(t/2). \tag{7.6.14}$$

By passing to limit $t \to 0$ one can verify that (7.6.14) is also valid for $t = 0$. By consecutive application of (7.6.14) we obtain the relation

$$\Lambda(t) = \prod_{k=1}^{n} \frac{\sin(t/2^k)}{(t/2^k)} \cdot \Lambda(t/2^n). \tag{7.6.15}$$

Next, from (7.6.15), (7.6.11) and the elementary inequality $|\sin \tau| \leq |\tau|$, relation (7.6.13) follows and thus the lemma is proved.


**Corollary 7.6.1**   *For every $t \in \mathbb{R}$*

$$\Lambda(t) = \prod_{k=1}^{\infty} \frac{\sin t2^{-k}}{t2^{-k}} \tag{7.6.16}$$

*that is*

$$\int_{-1}^{1} \cos(tx) \cdot \lambda(x)\,dx = \prod_{k=1}^{\infty} \frac{\sin t2^{-k}}{t2^{-k}}.$$


From Lemma 7.6.2 it has been proved that:
(a)  the infinite product 7.6.16) converges (see, e.g. [55], p 407);
(b)  the value of this product [50], [51] can be found;
(c)  the speed of the approximation, the estimate (7.6.13), which can be used for tabulating of the function $\Lambda(t)$, can be found.

The interesting question for the location of the zeros of the analytic function (7.6.9) remains open.

Further, (7.6.10) and (7.6.16) imply the following integral representation of the function $\lambda(x)$:

$$\lambda(x) = \frac{1}{2\pi} \int_{-\infty}^{+\infty} \cos(tx) \cdot \Lambda(t)\,dt$$

$$= \frac{1}{2\pi} \int_{-\infty}^{+\infty} \cos(tx) \cdot \prod_{k=1}^{\infty} \frac{\sin(t2^{-k})}{t \cdot 2^{-k}}\,dt$$

$$= \frac{1}{\pi} \int_{0}^{+\infty} \cos(tx) \cdot \prod_{k=1}^{\infty} \frac{\sin(t2^{-k})}{t \cdot 2^{-k}}\,dt. \tag{7.6.17}$$

Formula (7.6.17) shows that the function $\lambda(x)$ is the atomar function $\mathrm{up}(x)$ of Rvachev and Rvachev [50] (see [51], p 109).

In this way, on the basis of the quasi-splines the theory of the atomar function is developed. This function plays an essential role in non-classical methods of the theory of approximation for boundary value problems (see [51], [56]). Several new properties are found. First, let us mention, that it is the unique solution of Problem C. Next, these are the properties stated in Theorems 7.3.4, 7.3.5, 7.3.7, 7.3.8, 7.3.9, 7.5.1 etc., Corollaries 7.3.2, 7.3.4, Lemma 7.3.5 etc. Lemma 7.3.1, asserts that the atomar function coincides with each of its basic quasi-spline. This means that in the basis of the fundamental rational quasi-splines one may put the atomar function for every $s \in \mathbb{N}_0$ as

$$b(x) = \frac{\lambda^s(x)}{\lambda^s(x) + \lambda^s(1-x)}, \quad x \in [0, 1], \tag{7.6.18}$$

and then all the investigations, made in Chapter 6 remain valid. Taking into account that if $b(x) \in A_s$, then $b(x) \in A_r$ for every non-negative integer $r \leqq s$. For practical aims we can take a large enough $s = s_0$ and tabulate the function (7.6.18) with a step small enough in the segment $[0, 1]$. Thus we can obtain a practically universal approximation table. We may use it according to the manner developed in Chapter 6 for approximation of functions of a large enough class without looking for a system of approximating aggregates for every specific problem, as is the case for splines.

Let us note explicitly that we will keep the notation $\lambda(x)$.

The atomar function is non-negative, continuous, infinitely differentiable and satisfies the condition

$$\int_{-\infty}^{+\infty} \lambda(x)\,dx = \int_{-1}^{1} \lambda(x)\,dx = 1. \tag{7.6.19}$$

Therefore, it can serve as the density of a continuous random variable $X$. For such a random variable we say, that it has atomar distribution. The mathematical expectation, variance and standard deviation are respectively

$$MX = 0, \quad DX = \mu_2 = \frac{1}{9}, \quad \sigma = \frac{1}{3}. \tag{7.6.20}$$

Let us evaluate the central moments of the atomar function:

$$\mu_k \stackrel{\mathrm{def}}{=} \int_{-\infty}^{+\infty} x^k \cdot \lambda(x)\,dx = \int_{-1}^{1} x^k \cdot \lambda(x)\,dx, \quad k \in \mathbb{N}_0. \tag{7.6.20}$$

Then (7.6.19)–(7.6.21) imply

$$\mu_0 = 1, \quad \mu_1 = 0, \quad \mu_2 = \frac{1}{9}, \quad \mu_{2k-1} = 0 \quad (k \in \mathbb{N}). \tag{7.6.22}$$

It remains to evaluate the even central moments $\mu_{2k}$ $(k \in \mathbb{N})$. We have

$$\mu_{2k} = \int_{-1}^{1} x^{2k} \cdot \lambda(x) dx$$

$$= \int_{-1}^{1} x^{2k} \cdot \int_{2x-1}^{2x+1} \lambda(u) du \, dx$$

$$= \int_{-1}^{1} \int_{(u-1)/2}^{(u+1)/2} x^{2k} \cdot \lambda(u) dx \, du$$

$$= \int_{-1}^{1} \lambda(u) \left. \frac{x^{2k+1}}{2k+1} \right|_{(u-1)/2}^{(u+1)/2} du$$

$$= \frac{1}{(2k+1) \cdot 2^{2k+1}} \cdot \int_{-1}^{1} \lambda(u)[(u+1)^{2k+1} - (u-1)^{2k+1}] du$$

$$= \frac{1}{(2k+1) \cdot 2^{2k+1}} \cdot 2 \cdot \sum_{s=0}^{2k}{}' \binom{2k+1}{s} \cdot \int_{-1}^{1} u^s \cdot \lambda(u) du$$

$$= \frac{\mu_{2k}}{2^{2k}} + \frac{1}{(2k+1) \cdot 2^{2k}} \cdot \sum_{s=0}^{2k-2}{}' \binom{2k+1}{s} \cdot \mu_s.$$

Thus,

$$\mu_{2k} = \frac{1}{(2k+1) \cdot (2^{k+1} - 1)} \cdot \sum_{s=0}^{2k-2}{}' \binom{2k+1}{s} \cdot \mu_s, \quad k \in \mathbb{N} \tag{7.6.23}$$

where $\sum'$ means that the summing is spread on the even $s$ only.

Recurrence relation (7.6.23) and equations (7.6.22) allow us to evaluate consecutively all desired central moments of the random variable $X$ with the atomar distribution. For example,

$$\mu_4 = 0.028(148)$$

$$\mu_6 = 0.0097925589989\ldots$$

$$\mu_8 = 0.0040824582233\ldots$$

etc.

Let $X_n$, $n \in \mathbb{N}$ be independent random variables, uniformly distributed in the corresponding intervals $[-1/2^n, 1/2^n]$, $n \in \mathbb{N}$. Let us denote by $f_n(x)$ the density of $X_n$, i.e.

$$f_n(x) = \begin{cases} 0, & x \notin [-1/2^n, 1/2^n] \\ 2^{n-1}, & x \in [-1/2^n, 1/2^n] \end{cases}.$$

The characteristic function [57] of $X_n$ is

$$g_n(t) = \int_{-\infty}^{+\infty} e^{itx} \cdot f_n(x)\,dx = 2^{n-1} \cdot \int_{-2^{-n}}^{2^{-n}} \cos(tx)\,dx = \frac{\sin(t2^{-n})}{t2^{-n}}$$

that is

$$g_n(t) = \frac{\sin(t2^{-n})}{t2^{-n}}, \quad n \in \mathbb{N}. \tag{7.6.24}$$

We consider the series

$$\sum_{n=1}^{\infty} X_n \tag{7.6.25}$$

Since $|X_n| \leqq 2^{-n}$ and the series is convergent, then series (7.6.25) is absolutely (and uniformly) convergent. Denoting

$$X = \sum_{n=1}^{\infty} X_n \tag{7.6.26}$$

then (see [57]) the characteristic function $g(t)$ of $X$ is the product of the characteristic functions of the random variables $X_n (n = 1, 2, \ldots)$:

$$g(t) = \prod_{n=1}^{\infty} g_n(t) = \prod_{n=1}^{\infty} \frac{\sin(t2^{-n})}{t2^{-n}} = \Lambda(t). \tag{7.6.27}$$

Hence, the random variable (7.6.26) has the atomar distribution. In such a way, the following proposition is proved.

### Theorem 7.6.1

*If the random variable $X$ is a sum of independent random variables $X_n$ ($n = 1, 2, \ldots$), uniformly distributed in the intervals $[-2^{-n}, 2^{-n}]$ ($n = 1, 2, \ldots$), respectively, then it has the atomar distribution.*

# References


[1] Nikol'skii S M 1979 *Quadrature formulas* (Moscow: Nauka) (Russian)

[2] Sendov B and Popov V 1983 *Averaged moduli of smoothness.* (Sofia: Bulg. Acad. Sci.)

[3] Sendov B 1967 *Approximation with respect to Hausdorff's distance*: *DSc Thesis* (Moscow: Steklov Mathematical Institute) (Russian)

[4] Korovkin P P 1969 An attempt at an axiomatic construction of some questions of the approximation theory. *Ouchebnie Zapiski Kalininskogo Pedagogicheskogo Instituta* **69** 91–109 (Russian)

[5] Dolzhenko E P and Sevast'anov E A 1976 On the approximation of functions in Hausdorff's metrics by piece-wise monotonous (in particular, by rational) functions. *Mat. Sbornik* **101** 508–41 (Russian)

[6] Korneichuk N P 1976 *Extremal problems of the theory of approximations* (Moscow: Nauka) (Russian)

[7] Natanson I P 1949 Constructive theory of functions (Moscow–Leningrad: Gostechizdat) (Russian)

[8] Smolyak S A 1965 *On the optimal recovery of functions and of functionals of them*: *PhD Thesis* (Moscow: Moscow State University) (Russian)

[9] Bahvalov N S 1971 On the optimality of linear methods of approximation of operators on the convex classes of functions *J. Vich. Mat. i Mat. Fiz.* **11** 4 1014–18

[10] Boyanov B 1978 *Methods for approximate calculation of integrals* (Sofia: Nauka i Izkustvo) (Bulgarian)

[11] Nikol'skii S M 1946 Approximation of functions by Trigonometric polynomials in the mean. *Izv. Acad. Nauk SSSR, Ser. Mat.* **10** 207–56

[12] Korneichuk N P 1984 *Splines in the approximation theory* (Moscow: Nauka) (Russian)

[13] Favard J 1936 Application de la formule sommatoire d'Euler á la demonstration de quelques proprietes extremales des integrals des fonctions periodiques. *Math. Tidskrift* **4** 81–94

[14] Kolmogorov A N 1930 On inequalities between the upper bound of consecutive derivatives of functions on infinite interval. *Ouchenie Zapiski MGU* **30** 3–16 (Russian)

[15] Sard A 1949 Best approximative integration formulas. *Am. J. Math.* **71** 80–91

[16] Sard A 1963 Linear approximation. *Math. Surveys No. 9* (Providence, RI: American Mathematical Society)

[17] Boyanov B 1980 Optimal recovery of functions and functionals. *Summary of DSc Thesis* Sofia (Bulgarian)

[18] Turezkii A H 1951 On estimates of approximations by quadrature formulas for functions satisfying Livschitz's condition. *Ouspehi Mat. Nauk* **6** 5 166–71 (Russian)

[19] Korneichuk N P 1968 Best cubature formulas for some function classes of several variables. *Mat. Zametki* **3** 5 565–76 (Russian)

[20] Proinov P 1980 A remark on the convergence of a general quadrature procedure with positive weights. *Proc. Conf. Constructive Function Theory* (Blagoevgrad) 1977 Sofia pp 121–5 (Russian)

[21] Sobol' I M 1969 Multi-dimensional quadrature formulas and Haar functions (Moscow: Nauka) (Russian)

[22] Van der Corput J G 1935 Verteilungsfunktionen. *Proc. Kon. Akad. Wetensch.* (Amsterdam) **38** pp 264–9

[23] Roth K F 1954 On irregularities of distribution. *Mathemathika* **1** 2 73–9

[24 Kirov G On the optimal recovery of functions I. *Godishnik na VUZ, Priložna matematika* (to appear)

[25] Kirov G On the optimal recovery of functions II. *Priložna matematika* (to appear)

[26] Kirov G 1985 Optimal quadrature formulas on the classes $W^r H^\omega[0,1]$. *Compt. Rend. Acad. Bulg. Sci.* **38** 1 47–50 (Russian)

[27] Krylov V I 1967 Approximate calculation of integrals (Moscow: Nauka) (Russian)

[28] Kirov G H 1984 Optimal quadrature formulae for function classes $W^r H^\omega$. *Constructive theory of functions '84* (Sofia) 451–6

[29] Mysovskikh I P 1981 *Interpolatory cubature formulae* (Moscow: Nauka) (Russian)

[30] Kirov G H 1984 Optimal cubature formulae with constraints on the classes $H^{\omega_1 \omega_2}(D_2)$ and $H^\omega(D_2)$. *Compt. Rend. Acad. Bulg. Sci.* **37** **12** 1617–20 (Russian)

[31] Niederreiter H 1972 Methods for estimating discrepancy. *Proc. Symp. Applications of Number Theory to Numerical analysis* (Montreal) 1971 (New York: Academic)

[32] Proinov P and Kirov G 1983 On a quadrature formula for approximate integration of functions of the class $C^r[0,1]$. *Compt. Rend. Acad. Bulg. Sci.* **36** **8** 1027–30 (Russian)

[33] Proinov P and Kirov G 1984 Application of uniformly distributed matrices for approximation of functions and for numerical integration. *Compt. Rend. Acad. Bulg. Sci.* **37** **12** 1625–8 (Russian)

[34] Christov V 1983 A remark on the convergence of a quadrature procedure. *Compt. Rend. Acad. Bulg. Sci.* **36** **9** 1151–4 (Russian)

[35] Totkov G and Bazelkov M 1984 On the convergence of some quadrature procedures. *Compt. Rend. Acad. Bulg. Sci.* **37** **12** 1621–4 (Russian)

[36] Kirov G 1978 Approximation of functions by $k$-splines. In *Mathematics and Mathematical Education* (Sofia: Bulg. Acad. Sci.) 361–8 (Bulgarian)

[37] Rvachev V A 1973 Representation of splines by finite functions. *Dokladi cad. Nauk Ukrain. SSR, Ser. A* **2** 123–6 (Russian)

[38] Konovalov V N 1978 An approach to approximation of continuous functions on a segment by splines. In *Problems of the theory of approximation of functions and its applications* (Kiev: Ukrainian Academy of Sciences) pp 124–30 (Russian)

[39] Storchai V F and Ligun A A 1974 On the deviation of some interpolatory splines in the metrics $C$ and $L_p$. In *Theory of approximation of functions and its applications* (Kiev: Ukrainian Academy of Sciences) 148–57 (Russian)

[40] Velikin V L 1970 On the best approximation by spline-functions on classes of continuous functions. *Mat. Zametki* **8** **1** 41–6 (Russian)

[41] Kirov G 1982 Approximation of functions by rational quasi-splines. *Godishnik na VUZ. Priložna matematika* **18** **2** 41–56 (Bulgarian)

[42] Kirov G 1984 Approximation of functions by rational quasi-splines in the metric of $L_p[0,1]$. *Compt. Rend. Acad. Bulg. Sci.* **37** **11** 1455–8 (Bulgarian)

[43] Kirov G 1979 Exact estimates of approximations of differentiable functions by $k$-splines. *Compt. Rend. Acad. Bulg. Sci.* **32** **11** 871–4 (Russian)

[44] Kirov G and Proinov P 1983 On the approximation of functions of the class $W^r H^\omega[0,1]$ by quasi-splines. In *Constructive function Theory '81* (Sofia: Bulg. Acad. Sci.) pp 87–92

[45] Totkov G, Kirov G and Bazelkov M 1982 Approximation of functions of the class $W_p^r[0,1]$ by rational interpolatory splines. *Godishnik na VUZ. Priložna matematika* **18** **3** 33–9 (Bulgarian)

[46] Efimov A V 1961 Linear methods of approximation of some periodic functions *Matem. Sbornik* **54** **(96:1)** 51–90

[47] Ivanov K G Direct and converse theorems for the best algebraic approximation in $C[-1,1]$ and $L_p[-1,1]$ *Compt. Rend. Acad. Bulg. Sci.* **33** **10** (1980) 1309

[48] Ivanov K G 1983 A constructive characteristic of the best algebraic approximation in $L_p[-1,1]$ $(1 \leq p \leq \infty)$. In *Constructive function Theory '81* (Sofia) 357–67

[49] Ivanov K G 1982 On a new characteristic of function I *Serdica* **8** 262–79

[50] Rvachev V L and Rvachev V A 1971 On a finite function. *Dokladi Acad. Nauk Ukrain. SSR Ser. A* No. 8 705–7 (Ukrainian)

[51] Rvachev V L and Rvachev V A 1974 *Non-classical methods of the approximation theory in boundary value problems* (Kiev: Naukova Dumka) (Russian)

[52] Kirov G 1982 A representation of the derivatives of the basic quasi-splines and decomposition of polynomials. In *Mathematics and Mathematical Education* (Sofia: Bulg. Acad. Sci.) 193–9

[53] Kirov G and Totkov G 1981 On the location of the zeros of the derivatives of the function $\lambda(x)$. *Godishnik na VUZ. Priložna Matematika*

[54] Kirov G, Akteryan S G, Semerdžiev Kh and Popova L I 1980 Tabulating of the function $\lambda(x)$ (Collection of lectures on the yubileum session scientific session of VNVVU 'G Benkovski') *D. Mitropolia* 36–40 (Bulgarian)

[55] Fihtengol'z G M 1948 Treatise on differential and integral calculus, vol. 2 (Moscow–Leningrad: Fizmatgiz) (Russian)

[56] Rvachev V L 1982 *Theory of the R-Functions and some of its applications* (Kiev: Naukova Dumka) (Russian)

[57] Lucacz E 1979 *Characteristic functions* (Moscow: Nauka) (Russian)

[58] Proinov P 1981 Uniformly distributed sequences and approximate integration of functions. *Summary of a PhD Thesis* (Sofia)

[59] Malozemov, V N 1967 On the deviation of a polygonal line. *Mat. Zametki* **1 5** 537–40

[60] Storchai V F 1969 On the deviation of polygonal lines in the metrics of $L_p$ *Mat. Zametki* **5 1** 31–7 (Russian)

[61] Loginov A S 1969 Approximation of continuous functions by polygonal lines *Mat. Zametki* **6 2** 149–60

[62] Rvachev V L Rvachev V A 1972 On the representation of finite functions by polynomials. *Matem. Fiz.* **11** 126–9 (Russian)

[63] Rvachev V L and Rvachev V A 1975 Atomar functions in mathematical physics. In *Mathematization of knowledge and science–technological progress* (Kiev: Ukrainian Academy of Sciences), 188–99 (Russian)

[64] Kirov G On the optimal recovery of functions III. *Godishnik na VUZ. Prilozna matematika* (Bulgarian) (to appear)

[65] Storchai V F 1972 Approximation of continuous functions of two variables by spline-functions in $C$-metrics. In *Issledovaniya po sovremennim proplemam summirovaniya i priblizenie funczii i ih prilozenia* (Dnepropetrovsk) 66–8 (Russian)

[66] Storchai V F 1973 Approximation of functions of two variables by polytopial functions in uniform metrics. *Izv. VUZ-ov Matematika No. 8* 84–8 (Russian)

[67] Storchai V F 1975 Approximation of continuous functions of two variables bypolytopial functions and by spline-functions in uniform metrics. In *Issledovaiya po sovremennim problemam summirovaniya i problizeniya funkzii i ih prilozeniya* (Dnepropetrovsk) 82–8 (Russian)

[68] Storchai V F 1976 Approximation of continuous functions of two variables by some interpolatory splines in $C$-metrics. In *Voprosi teorii priblizeniya funkzii i ee prilozenii* (Kiev) 186–93 (Russian)

[69] Malozemov V N 1966 On the deviation of polygonal lines. *Vestnik LGU* No 7 150–3 (Russian)

[70] Zavyalov Yu S, Kvasov B I and Miroshnichenko V L 1980 Methods of spline-functions. (Moscow: Nauka) (Russian)

[71] Proinov P D 1984 Uniformly distributed matrices and numerical integration. In *Constructive theory of functions '84* (Sofia) 704–9

[72] Kirov G and Totkov G 1982 On the location of the zeros of the function $\lambda(x)$. In *Differential equations and applications Rousse '81* (Rousse) 341–4 (Russian)

[73] Polya G and Szegö G 1954 *Aufgabe und Lehrsäatze aus der Analysis* Bd. 1, Aufgabe 75 (Berlin: Springer)

# Author Index

# Subject Index